SERVICE MODELLING

SERVICE MODELLING

Principles and Applications

Vilho Räisänen
Nokia Networks, Finland

Telephone (+44) 1243 779777

Email (for orders and customer service enquiries): cs-books@wiley.co.uk
Visit our Home Page on www.wiley.com

Other Wiley Editorial Offices

John Wiley & Sons Inc., 111 River Street, Hoboken, NJ 07030, USA

Jossey-Bass, 989 Market Street, San Francisco, CA 94103-1741, USA

Wiley VCH Verlag GmbH, Boschstr. 12, D-69469 Weinheim, Germany

John Wiley & Sons Australia Ltd, 42 McDougall Street, Milton, Queensland 4064, Australia

John Wiley & Sons (Asia) Pte Ltd, 2 Clementi Loop #02-01, Jin Xing Distripark, Singapore 129809

John Wiley & Sons Canada Ltd, 22 Worcester Road, Etobicoke, Ontario, Canada M9W 1L1

Wiley also publishes its books in a variety of electronic formats. Some content that appears in print may not be available in electronic books.

British Library Cataloguing in Publication Data

A catalogue record for this book is available from the British Library

ISBN-13: 978-0-470-01807-1
ISBN-10: 0-470-01807-0

Typeset in 10/12 Times by Laserwords Private Limited, Chennai, India
Printed and bound in Great Britain by Antony Rowe Ltd, Chippenham, Wiltshire
This book is printed on acid-free paper responsibly manufactured from sustainable forestry in which at least two trees are planted for each one used for paper production.

Contents

Acknowledgements

Firstly, thanks are due to Nokia corporation for the possibility to participate in activities that have contributed to this book.

The author would like to thank the following persons (in alphabetical order) for commenting on the different parts of the manuscript: Paul Hendriks, Jenny Huang, Mika Klemettinen, and Veli Kokkonen. These contributors, as well as an anonymous reviewer, are thanked for positive contribution to the clarity of the text. The author assumes responsibility for any possible shortcomings in incorporating their comments into the final version.

The author would like to acknowledge the technical discussions within Service Framework Team of TeleManagement Forum (TMF), the Architecture work package of EU FP6 MobiLife project, and the architecture work package of Wireless Word Research Forum (WWRF). Reading of the various documents of TMF relating to NGOSS (New Generation OSS), SID (Shared Information/Data[model]), and eTOM (Enhanced Telecom operations map) activities has provided valuable insight into the topic area of this book.

Being able to represent knowledge in a useful context is crucially important for a book having a relatively abstract topic like this one. To the extent that this goal has been reached, great part of the credit goes to exchange of thoughts with persons and organisations which cannot be named here, and who have provided valuable insight into what is important in the topic area. Within Nokia, discussions with Kimmo Raatikainen of Nokia Research Center and Ulla Koivukoski of Nokia Networks are acknowledged with gratitude.

This book was almost entirely written using LaTeX, ArgoUML, XFig, and Dia running on GNU/Linux system. Mozilla Firefox™ and OpenSSH provided necessary connectivity.

Preface

The importance of services is increasing. In Europe and the United States, an increasing part of the labour is employed in services sector. Liberalisation of services in general is an actively discussed topic within the European Union.

Digital services have reached sufficient maturity to achieve mass-market acceptance, and consequentially, also an increasing number of providers. As a consequence of these developments, effective ways of providing services is becoming increasingly important. The situation can be likened to improvements in manufacturing industry, resulting in higher productivity for complex products such as automobiles. Indeed, multiple references have been made to 'service science' lately.

The topic of this book is modelling of digital services. We start from near history, and proceed to description of state-of-the-art technologies and trends. Technology, business, and end-user perspectives are used to analyse developments. In addition to these, selected research projects are tapped for further input to service modelling, and framework for utilising service modelling is described.

In an attempt to be as generic as possible, multi-provider value nets are used in examples of the book. Same methods can nevertheless be used also for modelling services within individual stakeholders.

Vilho Räisänen

Abbreviations

3G	Third Generation
3GPP	Third Generation Partnership Project
3NF	Third Normal Form
5NF	Fifth Normal Form
ADSL	Asymmetric Digital Subscriber Line
AF	Assured Forwarding
AI	Artificial Intelligence
ASIC	Application-Specific Integrated Circuit
B3G	Beyond 3G
BCP	Best Current Practices
BE	Best Effort
BNF	Backus-Naur Form
BPEL[4WS]	Business Process Execution Language [for Web Services]
BPMI	Business Process Management Initiative
BPMN	Business Process Modelling Notation
BPSM	Business Process Semantic Model
BSC	Base Station Controller
BSS	Business Support System
BTS	Base Transceiver Station
CBE	Component-Based Engineering
CBE	Core Business Entity (OSS/J)
CFS	Customer-Facing Service
CIM	Computation Independent Model
CORBA	Common Object Request Broker Architecture
COTS	Commercial, Off-The-Shelf
CPU	Central Processing Unit
CRM	Customer Relationship Management
DiffServ	Differentiated Services
DAML	DARPA Agent Markup Language
DARPA	Defence Advanced Research Projects Agency
DEN	Directory Enabled Networking
DEN-ng	DEN new generation
DMTF	Distributed Management Task Force

DNS	Domain Name System
DoS	Denial of Service
DRM	Digital Rights Management
DSCP	DiffServ Code Point
DVB-H	Digital Video Broadcasting for Handhelds
EBNF	Extended BNF
EDW	Enterprise Data Warehousing
EF	Expedited Forwarding
ER	Entity/Relationship (model)
eTOM	Enhanced Telecom Operations Map
EU	European Union
FCC	Federal Communications Commission
FMC	Fixed-Mobile Convergence
FP6	Sixth Framework Programme
GAA	Generic Authentication Architecture
GGSN	GPRS Gateway Support Node
GPRS	General Packet Radio Services
GUI	Graphical User Interface
HLR	Home Location Register
HSDPA	High-Speed Downlink Packet Access
HSPA	High-Speed Packet Access
HSUPA	High-Speed Uplink Packet Access
HTTP	HyperText Transfer Protocol
IDL	Interface Description Language
IdP	Identity Provider
IEEE	Institute for Electrical and Electronics Engineers
IETF	Internet Engineering Task Force
IM	Instant Messaging
IP	Internet Protocol
IPv6	IP version 6
IMS	IP Multimedia Subsystem
IMSI	International Mobile Subscriber Identity
IPPM	IP Performance Measurement
ISO	International Standardisation Organisation
IT	Information Technology
itSMF	IT Service Management Forum
ITIL	IT Infrastructure Library
ITU	International Telecommunications Union
KQI	Key Quality Indicator
KPI	Key Performance Indicator
LAN	Local Area Network
LBS	Location-Based Services
MDA	Model Driven Architecture
MPLS	Multi-Protocol Label Switching
MVC	Model/View/Controller
MVNO	Mobile Virtual Network Operator

N/A	Not Applicable
NAT	Network Address Translation
NF	Normal Form
NFC	Near Field Communications
NGN	Next Generation Networks
NGOSS	New Generation OSS
Node B	base station in WCDMA networks
OBSAI	Open Base Station Architecture Initiative
OCL	Object Constraint Language
OMA	Open Mobile Alliance
OMG	Object Management Group
OOM	Object-Oriented Modelling
OOP	Object-Oriented Programming
ORB	Object Request Broker
OSE	OMA Service Environment
OSI	Open Systems Interconnect
OSPE	OMA Service Provider Environment
OSS	Operations Support System
OSS/J	OSS through Java (TM)
OWL	Web Ontology Language
P2P	Peer-to-Peer
PBM	Policy-Based Management
PDB	Per-Domain Behaviour
PDF	Policy Decision Function
PDM	Product Data Modelling
PDP	Packet Data Protocol
PIN	Personal Identification Number
PLM	Product Life cycle Management
PLMN	Public Land Mobile Network
PMNO	Physical Mobile Network Operator
PoP	Point of Presence
PSTN	Public Switched Telephony Network
PTT	Push-to-Talk
QoS	Quality of Service
R6	Release 6
RDF	Resource Description Framework
RFS	Resource-Facing Service
RMI	Remote Method Invocation
RNC	Radio Network Controller
RTCP	Real-Time Control Protocol
SAP	Service Access Point
SDP	Session Description Protocol
SDR	Software Defined Radio
SDU	Service Data Unit
SFT	Service Framework Team
SGSN	Service Gateway GPRS Support Node

SID	Shared Information/Data [model]
SIP	Session Initiation Protocol
SLA	Service Level Agreement
SLO	Service Level Objective
SLS	Service Level Specification
SMF	Service Modelling Framework
SoA	Service-Oriented Architectures
SOAP	Simple Object Access Protocol
SoIP	Services over IP
SSE	Software Service Engineering
SSO	Single Sign-On
TCA	Traffic Conditioning Agreement
TCP	Transmission Control Protocol
TCS	Traffic Conditioning Specification
THP	Traffic Handling Priority
TMF	TeleManagement Forum
TNA	Technology-Neutral Architecture
UCD	User-Centred Design
UDDI	Universal Description, Discovery, and Integration [protocol]
UMA	Unlicensed Mobile Access
UML	Unified Modelling Language
UMTS	Universal Mobile Telephony System
UWB	Ultra-Wide Band
VoIP	Voice over IP
VPN	Virtual Private Network
W3C	World Wide Web Consortium
WCDMA	Wideband Code Division Multiple Access
WfMC	Workflow Management Coalition
WLAN	Wireless Local Area Network
WP	Work Package
WSDL	Web Service Description Language
WFQ	Weighted Fair Queuing
WS-I	Web Services Interoperability Organisation
WSI	Web Services Interfaces
WSMF	Web Service Modelling Framework
WSMT	Wireless Services Measurement Team
WWI	Wireless World Initiative
WWRF	Wireless World Research Forum
XMI	XML Metadata Interchange
XML	eXtensible Mark-up Language
XOR	exclusive 'or'
XP	Extreme Programming

How to Read This Book

This book is organised into four parts, each consisting of chapters. In what follows, we shall first describe each part, and proceed to describe their interdependencies.

Description of Parts

Part One sets the stage for the book, describing the issues service modelling seeks to address. It includes discussion about technological state of the art and trends, (Chapter 1), modelling paradigms and technologies (Chapter 2), and industrial initiatives (Chapter 3).

Part Two addresses actual service modelling. Formulation of service modelling proceeds via listing of requirements for service modelling (Chapter 4), an account of management framework (Chapter 5), description of the service framework within which service model is used as well as summary of requirements and characteristics of services (Chapter 6) to description of service modelling patterns (Chapter 7).

Part Three provides examples of use of service modelling via examples. Chapter 8 provides an example of applying service modelling to fixed Internet, and Chapter 9 to mobile networks. An example of modelling distributed network is provided in Chapter 10.

Part Four provides a summary of the central themes of the book and outlines relevant future issues.

How to read Parts One and Two

Part One provides a description of the assumptions on which the model presented in Part Two is built. The assumptions are also summarised at the beginning of Part Two. Part Three illustrates the use of modelling concepts. Reader familiar with the world of packet-based services may start with Part Two and refer back to Part One on information on specific topics.

Part One is mostly description of state of the art, and can be assumed to be familiar for some of the readers, at least partly. To facilitate easier navigation through Part One, selected key messages have been shown in boxes like this:

This Is A Crystallisation of The Surrounding Text.

A reader fairly knowledgeable of the topic area can review the boxes and concentrate on issues of specific interest. In all the chapters, highlights of contents are summarised at the end.

Part Two has a specific structure to it, starting with requirements for service modelling. They are formulated based on summary of the background in Part One. Management framework is described next, constituting a description of the processes which make use of service modelling. A further building block is provided by service framework, which provides a description of services and facilitates description of services in terms of service quality and security. Examples of applying service framework to typical end-user services are provided here. The necessary groundwork being laid down by preceding chapters, service model is described using modelling patterns. Service modelling patterns make references to management framework and service framework.

Information About Trademarks and Copyrights

Disclaimer

List of Figures

List of Tables

Part I

Background

Scope of Part One

In order to understand service modelling, we need to account for the environment where it is used. In the first part of the book, we set the scene by providing a working definition for service modelling and describing the context for using it in Chapter 1. The context description includes a review of drivers for service management, as well as current perceptions of important trends in this area. Linking between service management and service modelling is made here.

Service modelling makes use of more general methods and paradigms. These approaches will be discussed in Chapter 2. We shall tie the discussion of the first two chapters together in Chapter 3, discussing relevant industry initiatives.

Part Two will build on the groundwork laid down in Part One, referring to the issues described below for defining a framework. This framework will then be used in the examples of Part Three.

1

Introduction

Service modelling, as a concept, is likely to be unknown to most technical people. After all, services have been with humankind for thousands of years, and trading concrete goods for services is indeed one of the hallmarks of a post-hunter-and-gatherer culture. One of the trends of our time is the growing importance of the production of services compared to the production of tangible goods such as automobiles or houses. Allowing ourselves to get a little philosophical for a while – but only for a while, don't worry – one could say that humankind directs its collective brainpower to issues that importantly need further work at a given moment. Production of mass-produced tangible goods being quite automated already, creation and operation of services has now got our attention. That's the philosophical part; practically oriented readers may breathe easier again!

One of the challenging areas of the day is the management of digital services. The challenge is not about managing them at all, but rather doing it in an optimal way. As we shall see in this chapter, management has thus far often been based on 'point solutions', leading to management frameworks and processes varying from service to service and from provider to provider. Our claim is that service modelling – understood in a broad sense and encompassing information and process modelling – makes a significant contribution to remedying the situation. In this chapter and later on in the book, the importance of service modelling is justified by the evolution in technology, business processes, and also the very paradigms that business models are based upon.

We shall define service modelling and explain why it is needed. In order to do the latter, we shall take a look at some trends within the industry and study some relevant technologies. Different approaches to modelling will be discussed in Chapter 2, and prior work will be summarised in Chapter 3.

1.1 Definition of Service Modelling

The service modelling framework will be described in the next part of this book, and there we shall present a definition of service modelling to be used within the context of this book. To get started, we shall formulate a 'working definition' of service modelling so that we can better assess its context and interactions with larger technological and business environments.

Service modelling concerns the representation of the relations between services and the resources supporting them on the one hand, and the relations between services and the way in which they are made available to the users of the services, on the other.

Before we can define service modelling in more detail, we need to define the entities and their inter-dependencies which we wish to study. The logical starting point is the concept of service itself. We shall start with a generic definition and move towards a definition of interest for us, observing the issues peculiar to the focused definition as we progress.

The most relevant definitions for 'service' in the Oxford English dictionary include the following (OED, 1995):

- The provision of a system of supplying a public need, e.g. transport, or supply of water, gas, electricity, telephone, etc
- The provision of what is necessary for the installation and maintenance of a machine or operation
- Assistance or advice given to customers after the sales of goods
- Supply[ing] [someone] with a service.

In the science of Economics, services are intangible commodities, whereas goods are *movable property or merchandise*, to again use a definition from OED (1995).

Based on the preceding definitions, it seems fair to say that service is a well-defined, immaterial entity which is provided by a party to other parties. When a service is something that is provided to customers in a contractual setting, the precise definition of the service becomes important. This is true irrespective of money changing hands in connection with service delivery or not.

The definition of a service may be contract-based, either directly between the service provider and the customer, or implicitly in the form of general terms of delivery. For example, in the former case, a contract is devised corresponding to the service to be provided, whereas in the latter case, the same implicit terms apply to a large number of customers. The two types of service definitions need not be mutually exclusive, but can co-exist. Separating multi-party aspects to a separate type of entity brings added clarity to analysis, as we shall see in the following text. In any case, the properties of services are related to terms of agreement provided to external parties.

A service typically needs resources to operate. Using the railway service as an example, it needs railway cars, engines, and tracks, for example. The difference between a resource and a service is important. A resource, by itself, is not a service. A network of railway tracks needs to be complemented by, for example, maintenance activities before it can be called *service* proper. Track maintenance is an example of service which is usually not directly visible to the end-user, but is used by another service (railway transport) which is directly visible to the end-user. Typically the provider of a service can provide different variants of the service by adjusting related parameters.

Even the service that is visible to the end-user cannot usually be provided to customers as such. Terms of delivery, pricing, and rules of usage need to be added to the technical definition of a service for this reason. They represent an organised view of the essentials of the service from the customer viewpoint. In the railway example, the operator of the railway transport may provide guarantees for trains operating on schedule. The railway operator also typically provides relatively stable pricing schemes for the railway service.

Based on the above, one can identify:

- Packaging of services for sale or other kinds of provision to customers
- Services having well-defined technical meaning
- Resources needed to operate the service.

The delineation between the first and the second types of entities seems to be quite clear-cut. The first one is related to the conditions of making service available to customers, whereas the second one is a technical entity. Note that a service can still be visible to end-users. Regarding the second and the third entity types, we shall see later in this book examples about an entity sometimes being perceived as belonging to both classes at the same time, depending on the viewpoint chosen.

Services provide controlled access to supporting resources for provision to users.

The above tripartition allows the treatment of each of the three fundamental entities as separate class of variables. Perhaps the most obvious possibility is that one can 'package' existing services in different ways by varying the terms of delivery associated with them. Railway transport operators may provide special prices for daily commuters, or for those who frequently travel between two cities.

There is also a second possibility. If a business opportunity is identified and resources are known, it is possible to pick existing services, re-parameterise them, or define new technical services so that they best bridge the gap between business and resources. This is obviously easier in some industries than others – in railway transport, for example, the possibilities of varying actual services are limited. For digital services, on the other hand, it seems that there is ample room for novel technical innovations.

Thirdly, given the business boundary conditions and services that are operated, needs and trends related to developing resources can be identified. Increase in traffic may give rise to the need to lay tracks on new routes, or install additional tracks on existing routes to extend capacity.

A note needs to be made here in passing about the multiplicities related to inter-relations between the classes of entities making up this trio. A service can obviously require multiple kinds of resources to operate, as illustrated in the railway example above. The same applies to the relation between things sold to customers and services, too: the commercial concept can aggregate multiple technical services. As we shall see, generally speaking, products can also aggregate other products, services can aggregate services, and resources can aggregate resources.

The discussion about the railway example suggests that it is important to understand how resources, services, and packages sold to end-users fit together. The provider of the railroad network infrastructure service needs to know which routes need special attention in maintenance, and how much traffic there is on particular routes. The operator of the actual transport service needs guarantees that the railroad network is in good enough condition. In addition, sufficient number of engines and railway cars must be available at specific locations at the right time. These, in turn, depend on the usage of the particular packages (first class, second class, special cars) there are on different routes.

We have now proceeded to the point where we are ready to provide the first definition of service modelling. It is as follows:

Service modelling amounts to the representation of relations between what is provided to customers, the technical definition of the services, and the resources needed for operating the service.

The above definition immediately gives rise to the question as to what kind of relations we are talking about here. To have a deeper insight into the issue at hand, we shall take a look at typical tasks in service modelling from different viewpoints, namely, those of service topology, stakeholders, processes, and data ownership. We shall discuss each of the viewpoints as a single entity for purposes of clarity, but it should be kept in mind that the actual implementation of each viewpoint may be actually distributed and consist of multiple systems.

Service topology

The preceding discussion has provided the first justification for service modelling: the ability to represent relations between items provided to customers, technical implementation of services, and resources needed because each of the three entity types can be viewed as being the 'free variable', depending on whether the purpose is to package technical services in different ways, identify technical services that best fulfill business needs, or identify resource development needs. Information about the inter-dependence of entities, called *service topology*, is needed to analyse the effects of changes to the arrangement.

A simple example of service topology is given in Figure 1.1. It seeks to illustrate the fact that individual resources can be used by multiple technical services, and individual

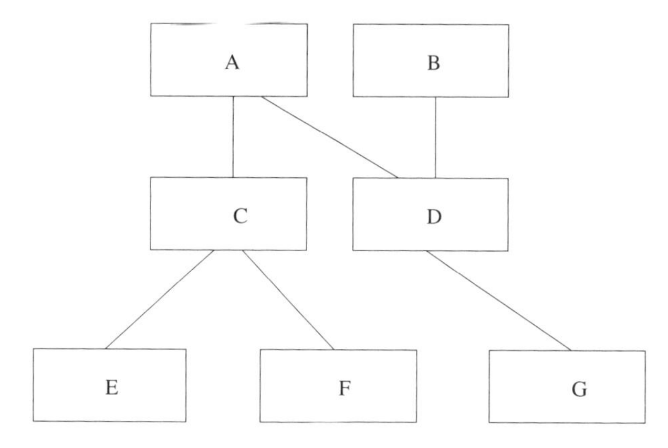

Figure 1.1 An example of service topology. Entities A and B are items provided to customers, and C and D are technical services. E, F, and G are resources

technical services by multiple items provided to customers. For simplicity, illustrations of relations between entities of the same type have been left out of the Figure.

The ability to represent dependencies between types of entities is crucially important for creating, operating, and managing services. As we shall see later in this book, the importance of service topology is highlighted by increasingly complex services and systems. During service creation, dependencies tell which kinds of resources need to be configured for the service to operate, and which end-user packagings depend on a particular service. In the operations phase, dependencies allow linking of performance (and non-performance) of individual resources to services and saleable objects. When services are modified or removed, dependencies again help in keeping track of which kinds of resources need to be configured and which saleable objects are affected.

Service topology represents static dependencies between resources, services, and provisioned entities.

A note relating to the use of service topology in connecting specific services is in order here. Service topology is usually expressed in terms of types of entities, e.g. saying that an entity of type 'A' requires services of types 'C' and 'D'. Making sure that sufficient resources are available for a particular instance of end-use packaging, an instance of 'A' requires the existence of the instances of the entities of types 'C' and 'D'. We shall return to the difference between types and instances in connection with modelling in Chapter 2.

Due to the viewpoint dependence of entity classification referred to above, different actors can represent the same environment – or a part thereof – with different service topologies. Thus service topology may be dependent on the viewpoint. We shall see later examples of multiple views being generated from a single service topology.

Stakeholders

In the railway example, we had the railway network service provider, the railway transport service provider, and customers listed as stakeholders. As we shall see below, the situation may be more complex and involve multiple providers. In such an environment, the benefits of having up-to-date information about inter-relations become further pronounced.

We shall next provide a more generic example of the types of stakeholders involved in service provision. Frameworks more directly related to the technical topic area of the book will be encountered later. The stakeholder framework for the present purpose is shown in Figure 1.2.

The stakeholders are as follows:

- End-user – the ultimate user of the service
- Subscriber – the party for whom service is provisioned. May be an employer of the end-user, for example
- Facilitator – provides access to services. Could be an owner of the railway station building for train service or mobile network operator for mobile packet-based services
- Service provider – provides well-defined services as provisioned entities to subscribers. Performs packaging of services for subscriber

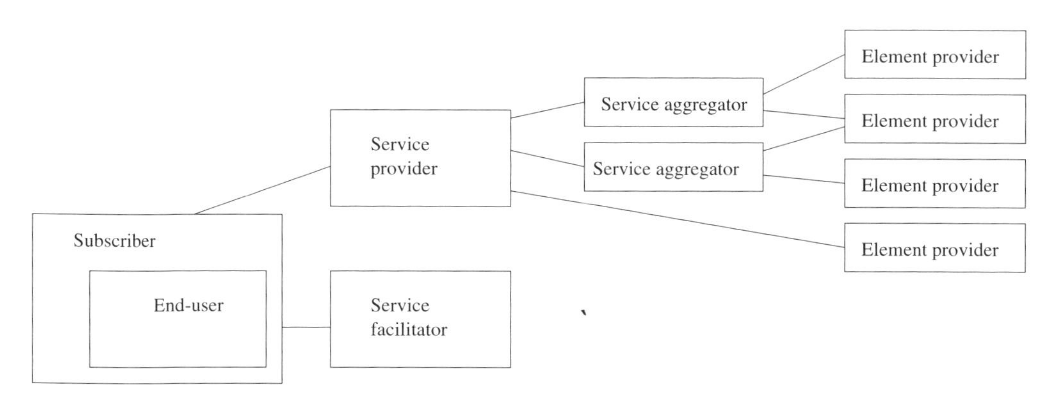

Figure 1.2 Stakeholders of service provisioning

- Aggregator – aggregates elementary building blocks of services. This stakeholder type is a subcontractor who buys some parts of services from other providers. Performs packaging of services for service provider
- Element provider – provides individual building blocks of services. Performs packaging of services for aggregator.

We can observe that different kinds of packaging of services – both business-to-customer and business-to-business – are involved.

Even though the stakeholder list above is better than simply having a service provider and a customer, it is not complete. We shall see some examples of more elaborate frameworks later in the book. The stakeholders typically have well-defined relations to the entities in service topology. Examples of this will be provided later.

Summarising the preceding discussion, the presence of different types of stakeholders leads to the requirement that stakeholder-specific views to service models need to be supported. Referring to the railway service once again, detailed information about the condition of the railroad network is relevant to the provider of the network service and the transport service. Information about schedules on a particular route is relevant to users of the service and the operator of the service.

Stakeholder analysis provides multiple viewpoints to service modelling.

Figure 1.2 represents a classification of the stakeholders and their inter-relations. It should be noted that the purpose of stakeholder diagrams is not to represent complete business models of the stakeholders but rather depict value nets on a higher level.

Processes

Management of services is closely related to processes, or the order in which different tasks required for creating and operating a service are carried out. In the following text, we shall discuss implications for service modelling from the viewpoint of process management and control. Management of processes is here viewed as defining the processes and control mechanisms to be used in it, whereas control relates to applying the mechanisms in the course of an instance of the process.

Taking packet-based service creation as an example, one can, generally speaking, identify generic phases in service creation (Koivukoski and Räisänen, 2005):

- Business analysis and business requirements definition
- Technical design
- Technical implementation
- Configuration and provisioning.

These phases will be discussed in more detail later on in a more generic context. As we shall see, one can delve deeper into operational management of services, for example.

The basic need for process management functionality is facilitating the dividing of a task into subtasks, and controlling the order in which individual subtasks are carried out, as well as tracking and back-tracking the process as a whole. The control functionality needs to support both 'normal flow' and 'corrective flow'. Referring to the simple service creation phasing example above, 'normal flow' would entail proceeding from the business phase to technical implementation in the order in which phases are listed above. 'Corrective flow', on the other hand, could mean returning to the technical design phase or even business requirements definition phase, based on findings in the technical implementation phase, for example. Corrective flow can also relate to retracing the sequence of subtasks within a larger task.

The processes relating to service management are typically specific to an enterprise, even though generic patterns can be identified on a sufficiently high abstraction level. Examples of this will be provided later in the book. A typical example of requirements of a process control system is the ability to track the status of an ongoing process flow and compare it to the planned schedule.

The ability to roll back a process or a part of it can be a requirement for process management. It enables moving back to a (at least partly) functional state in case the execution of the process is hindered.

Process control functionality must also support flexible allocation of individual parts of the process to members of management staff. The system should also support definition and supervision of timetables for each task.

Requirements for process management and control vary depending on the extent to which the 'production system' is affected. Railway tracks can be laid for new routes, often mostly independently of normal traffic, except when new tracks are connected to the existing network. Replacing faulty tram tracks in the heart of a city is a different kind of an example, with the extra boundary condition of attempting to minimise disruption to normal traffic.

Processes represent dynamic aspect of service modelling.

Summarising, service modelling should support defining and supervision of process flows, each consisting of a temporally ordered sequence of steps. Also, conditional branching of process flows should be supported. Service model should advantageously support reverting back to earlier phases of the process.

Information management

An important part of service management is making sure that right information is available in a timely manner in the correct format for the task at hand. This may include, for example, conveying information between different storage facilities and making data conversions or rendering of different representations of a particular piece of information.

Storage of information is an important topic in its own right – in large companies, the information needed by business management may need to be collected from multiple places, combined and processed in specific ways to provide necessary metrics. If infrastructure systems are provided by multiple vendors, questions of technical integration and ensuring data consistency and proper data conversions arise. Solutions for optimising information management may relate to storage locations and data formats.

Apart from obvious requirements arising from running on computers, such as support for backups, information management needs to support the definition and enactment of access rights to information. In a multi-provider environment, this should favourably support information management across corporate boundaries.

Permissions management, synchronisation, and representation formats are typical issues of relevance for information management.

Management of information should ideally link to service topology, stakeholders involved, and processes in such a way that it is an enabler for the other viewpoints, and overlapping work in different viewpoints is minimised.

Summary

We have defined service management as consisting of representation of inter-relations between what is sold to the customers, services as technical implementations, and resources supporting services. We have discussed representation of inter-relations from the viewpoints of service topologies, stakeholders, processes, and information management.

We have seen that service topology needs to be able to represent dependencies between entities and support instantiation for particular environments, and lend itself to be rendered to stakeholder-specific views. Description of temporal ordering of parts of processes needs to be supported, and the need to be able to revert to earlier phases of the process sets requirements for information management. Different representations of the service model need to be supported, and management of access rights to data is important for applicability to multi-stakeholder environment.

Up to now, the discussion about service modelling has been rather generic. We shall now start analysing the specific needs related to providing packet-based services.

1.2 Packet-based Services

In this section, we move into the realm of packet-based services by discussing some industry trends relevant to providing packet-based services. We shall describe both the state of the art and emerging trends for packet-based services, using a number of viewpoints. The summary of the state of the art and trends will be used as a background for other parts of this book, including a discussion about emerging technologies in Section 1.3, review of modelling in Chapter 2 and industry initiatives in Chapter 3. It is also useful background for service modelling requirements described in Chapter 4.

'Packet-based' refers to the delivery of digital content using packet-switched paradigm. Having 'packet-based' in the section title does not mean that all protocol stacks are packet-based. For example, in Generic Service Modelling (GSM)-based systems, it is possible to use either packet-switched General Packet Radio Services (GPRS) or circuit-switched data for using Wireless Application Protocol (WAP) services. Similarly, fixed Internet access can make use of both Asynchronous Digital Subscriber Line (ADSL) or Public Switched Telephone System (PSTN). In the subsequent technology examples, we shall mostly use GPRS and Universal Mobile Telephony System (UMTS), but that should not be interpreted as being restrictive from the viewpoint of applications in any way.

Packet-based services relate to transfer of digital content over different access technologies.

The International Telecommunications Union (ITU) has described global information infrastructure in (ITU-T Recommendation Y.110, 1998), providing a vision to frame our present discussion. The recommendation starts by saying that such an infrastructure should enable people to use communication services securely, using open application and all modes of communication, independently of location and time, at acceptable cost and quality. Global information infrastructure is viewed to have been born at the intersection of telecommunications, IT, and entertainment. The vision sounds good, all right, but the problem is, of course, getting there. Even though some aspects are already reality for mobile services, it is fair to say that there is still work to be done on this front. The approach of (ITU-T Recommendation Y.110, 1998), as well as some other activities discussed here, is to analyse the situation in terms of roles and models. The text of this section is to be viewed as an attempt to provide 'raw material' for such processing.

In what follows, an attempt is made to address both fixed Internet and mobile networks. We shall start with a summary of the state of the art, and then proceed to describe some relevant industry trends.

We concentrate on the Internet era and leave aside earlier technologies such as X.25.

1.2.1 State of the Art

We shall describe state of the art, using three viewing angles: technology, business models, and the end-user perspective.

Technology

Packet-based services were first provided to consumers on a large scale via fixed Internet access technologies, starting with technologies like 2BaseT, 5BaseT, and 10BaseT variants of Ethernet/802.3 in the corporate environment and PSTN. The spectrum of available technologies on the fixed access side has expanded to gigabite and 10G Ethernet and fibre-based technologies for corporates, and ADSL and fibre to homes for consumers, to name just the major trends.

Packet-based services are now also available while on the move, facilitated by systems like GPRS and UMTS, to use as example technologies that are based on the evolution of GSM. Third-generation mobile systems such as UMTS provide advanced multi-service capabilities and more advanced service quality support capabilities (Koodli and Puuskari, 2001; Laiho and Acker, 2005).

For both fixed and mobile access technologies, the most obvious feature resulting from this evolution is larger throughput available via a single link. Maximum theoretical downlink link layer throughput attainable for a single user in Wideband Code Division Multiple Access (WCDMA) variant of UMTS is 2 Mbit/s, which will be increased to around 10 Mbit/s with the adoption of the High-Speed Downlink Packet Access (HSDPA). Parallel enhancements are planned for uplink direction with High-Speed Uplink Packet Access (HSUPA). (The combination of HSDPA and HSUPA is known as High-Speed Packet Access, HSPA). These numbers compare favourably with ADSL performance. The wide-area managed performance of WCDMA + GPRS + HSPA can be supplemented with hotspot technologies such as 802.11 Wireless Local Area Network (WLAN).

Parallel to increase in single-user throughput, other trends have relevance to providing packet-based services. Connecting to Internet via dial-up is now automated. GPRS and UMTS handsets can automatically activate connection to Internet, and home users are using home gateways or automated dialing scripts. GPRS, UMTS, and ADSL can also cope with 'always-on' access to the Internet. Even though this feature seems self-evident nowadays, it is a powerful enabler for packet-based services. For example, 'push mail', or obtaining instant notification of new e-mail in corporate inbox, typically makes use of always-on connectivity.

Increase in throughput and always-on access are major technical developments in the use of digital services.

As a consequence of the above factors, more innovative services are available in addition to the traditional trio of e-mail, data upload/download, and Internet browsing. Instant Messaging (IM), presence, and Voice over IP (VoIP) are examples of services that are enabled by technological advances. Some of the services are available via almost any kind of Internet access, whereas others require more advanced solutions. VoIP is an example of the latter – to provide true telephone quality voice independently of the number of concurrent users sets requirements for service quality support within the network (Räisänen, 2003a). Innovation in the area of packet-based services has also produced 'lightweight' versions of services. Taking VoIP as an example, using 'walkie-talkie' (push-to-talk) mode in communication instead of a 'full-duplex' (Internet telephony) mode lowers the requirements for the network considerably. We shall discuss the requirements and characteristics of the services in more detail in Chapter 6.

From the point of view of the end-user, it is important that service performance is predictable and consistent (Bouch *et al.*, 2000). Typically, end-users are not interested in service quality parameters in most cases, whereby service quality allocation should be transparent from the viewpoint of an end-user. Technically, services have different requirements regarding network delivery, as well as different characteristics. Internet telephony requires low end-to-end latency and guaranteed minimum bandwidth, whereas for data transfer applications, it is the total download time that counts. Internet telephony is an example of a service which has inherent service quality requirements, whereas data transfer performance can usually be designed (Räisänen, 2003a). This leads to the requirement of the network provider needing to be able to allocate network resources so as to best cater for both types of services, while attempting to maximise revenue (Räisänen, 2004). Resource utilisation is a particularly important issues in radio networks operating on licensed frequencies, where spectral resources are costly.

Predictable service quality is important for an end-user. Constant service quality may be better for an end-user than variable higher quality.

In GPRS/UMTS networks, services are associated with a mobile bearer the properties of which can be provisioned according to subscriber profile, in addition to the type of the service. This makes it possible to transparently allocate differentiated service quality according to the type of service. In 3GPP static service provisioning architecture, bearer properties are conceptually managed separately from service provisioning (Laiho and Acker, 2005). In this scheme, services are associated with a Service Access Point (SAP), which determines the service quality used for connecting to the service. This scheme is suitable for client/server type applications. The provisioning scheme is illustrated in Figure 1.3.

The IP Multimedia Subsystem (IMS) has been developed by the Third Generation Partnership Project (3GPP) as a generic service platform for IP-based connectivity services (Poikselkä *et al.*, 2004). Real-time services of IMS may have session-specific parameters, whereby 3GPP architecture supports dynamic linking of IMS session properties to network resources. Where Figure 1.3 depicts static service quality provisioning, this dynamic link can be properly called dynamic linking of sessions to bearers, bearers meaning here access channels for service usage. IMS is an implementation of the Session Initiation Protocol (SIP) framework defined by Internet Engineering Task Force (IETF)

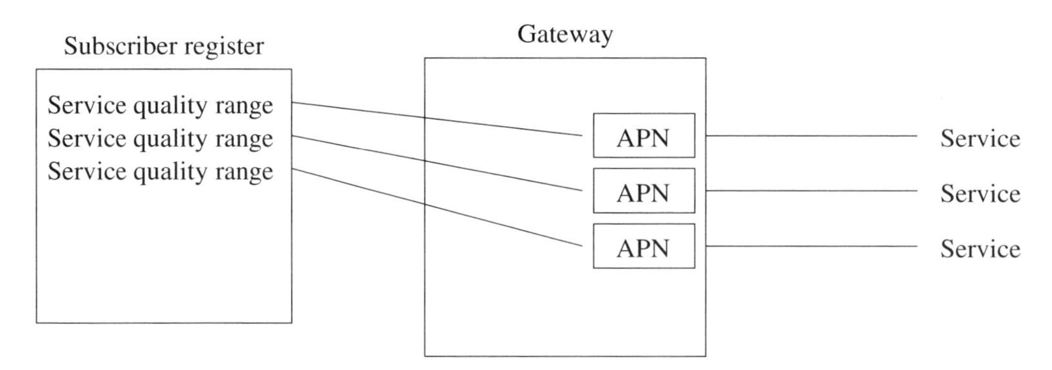

Figure 1.3 An illustration of the 'static' provisioning mode of 3GPP architecture. Subscriber's QoS profile defines service quality range for each Access Point Name (APN)

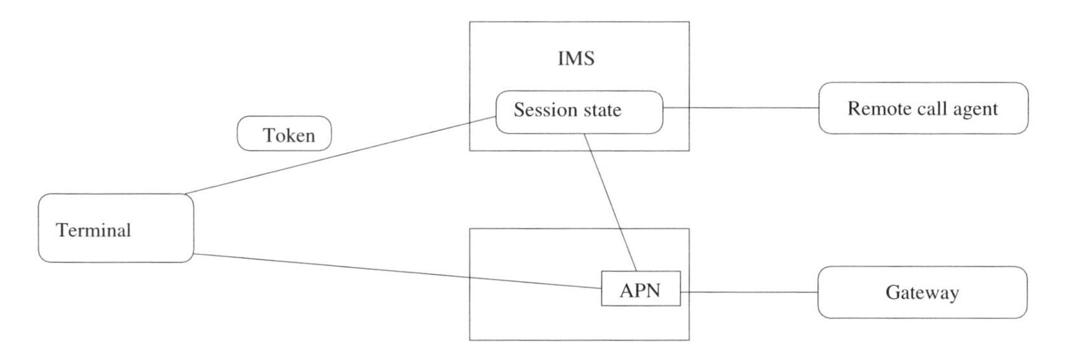

Figure 1.4 An illustration of the 'dynamic' provisioning mode of 3GPP architecture. IMS and terminal use a special token to correlate multimedia sessions with bearer properties

(Handley *et al.*, 1999). The SIP framework provides for reachability support and session management, and allows dynamic management and negotiation of media components during the session (Figure 1.4). IMS was designed to be a mobile-optimised implementation of the SIP framework. At the same time, it is a complete architecture for operating SIP-based applications. Work has been ongoing to generalise IMS for use in other access technologies as well. (Ahmavaara *et al.*, 2003). At the time of writing this, the general perception is that IMS will be widely used for both fixed and mobile networks and provide a platform for Fixed-Mobile Convergence (FMC).

3GPP mobile networks are used here for illustrating static and dynamic service provisioning paradigms.

More information about 3GPP service provisioning can be found in Appendix A.

The Internet Protocol (IP) provides a convergence layer, facilitating providing of services via multiple access technologies. This is a great opportunity for providing services across different access methods. The first steps of this development are already visible in the form of Virtual Private Network (VPN) clients, so that an employee equipped with a laptop can read e-mail and access corporate intranet via RJ45 cable in the office, via WLAN link in an airport hotspot, and by using handset as a modem in any location with cellular coverage. For services with more strict inherent service quality requirements, knowledge of characteristics of the access technology can be used for optimising service for a particular access technology (Koivukoski and Räisänen, 2005; Räisänen, 2004).

Internet Protocol is a convergence layer for digital services.

Technical capabilities for providing new services are relatively advanced at the moment, even if not all the advanced features are deployed in operational networks yet. The management of networks and services poses a challenge as such, given the increasing complexity of networking technologies and services. Additional constraints result from an increasing degree of competition between network and service providers. The situation has led to time and cost pressures, amounting to a requirement for management systems to operate effectively and accurately. Another consequence of the competitive landscape is the increasing interest toward the use of subcontracting in service provision. For example, a mobile operator may have a portal for subscribers containing news and weather content from specialised providers. Compared to the traditional way of working, this represents a change in business logic for participants in the end-to-end service provision value net.

In today's network management systems, Business Support Systems (BSS) and Operations Support Systems (OSS) are relatively separate from each other. A relatively new discipline entering the playfield of existing systems, the management of packet-based end-user services is in the process of causing a paradigm shift for the reason that it can be increasingly viewed as having aspects relating to both BSS and OSS. We shall return to this later in this book.

Operator systems are typically multi-vendor in nature, consisting of systems of different makes and models purchased over an extended period of time. Also, systems developed by operators themselves are used. From the system management viewpoint, different vendors' systems are often using disparate information models (Martin-Flatin *et al.*, 2003), resulting in a need to build the production system of an operator as a series of systems integration projects. This is clearly sub-optimal in view of striving for quick and accurate service provision, and has often led to long project duration in creating new services (Koivukoski and Räisänen, 2005). Disparate information models also make it difficult to create and manage end-user services and their performance levels.

Proprietary information models constitute a major challenge in multi-vendor integration today.

It is also not in the interest of either service providers or vendors to subscribe to this 'silo-world' paradigm in the long run. Today's systems also typically do not provide good enough support for multi-provider value nets to accommodate rapid service creation across corporate boundaries.

At the moment, access technologies are managed using dedicated management schemes. Not only are the management systems differ from vendor to vendor, but management processes and paradigms are also different in Information Technology (IT) systems and mobile networks. Nevertheless, on a sufficiently high level, processes within fixed network providers and within mobile operators can be described with common models. We shall touch upon this issue in Chapter 2.

Given the situation outlined above, there is clearly room for improvement. We shall next take a look at some topical industry trends.

Business models

Currently, Internet connectivity is purchased from an access network operator. The business relation between the party responsible for paying the bill – i.e., subscriber – and the access network provider can be either contract-based or purchased for the duration of access network use. For computer connectivity, examples of these variants include agreement between an enterprise and ADSL or fibre access provider on the one hand, and credit card payment for roaming WiFi hotspot access on the other, respectively. For mobile networks, pre-paid and post-paid subscriptions provide corresponding examples. The relation between subscriber and end-users may be many to one, which is the case for an enterprise paying the phone bills of employees.

In the fixed Internet, the subscriber typically has separate agreements with providers of packet-based services. End-users may have their own agreements with service providers in addition to the agreements made by the subscriber. It is also possible that a number of services are included in the price of the access. Decoupling of access and services provides greater flexibility, but requires separate agreements and charging settlements between subscriber and different providers.

In the mobile domain, there are advanced capabilities for charging for packet-based services. Subsequently, use of services is typically charged as a part of the mobile phone bill. This is convenient from the viewpoint of the subscriber. Alas, the current form of interfacing of access providers and service providers is relatively inflexible, typically supporting real-time information exchange about neither service usage nor charging information. The creation of new chargeable services currently requires a typically longish and relatively rigid process. From the viewpoint of service providers, the situation is made more challenging still by the fact that such arrangements are needed towards multiple access network providers.

During the last years, the emergence of access capacity resellers has been a major trend. Mobile Virtual Network Operators (MVNOs) have acquired a part of the mobile access market in some countries. From the viewpoint of business models, MVNOs have agreements with the (physical) network access provider and service providers.

Convergence of fixed and mobile networks on one hand and support for new value nets on the other require new ways of thinking.

The flexibility of Internet Protocol–based services is opening new possibilities also for services which have been traditionally provided over circuit-switched access. This also has an impact on the types of business stakeholders. We shall return to this in Section 1.2.2.

End-user perspective

In post-industrial countries, access to the Internet has become a commodity almost like water and electricity. Competition has lowered the prices, and provided multiple ways of connecting to the same information. The author has recent experience of the practical value of the latter; during a couple of days off in Vienna, Austria, it was convenient to be able to check the history of Urania on the spot in a café using GPRS access to a popular Internet search machine. Having access technology specific bills and service provider specific bills, on the other hand, is not particularly convenient.

The end-user is not typically interested in details of the access technologies. Aside from difference in throughput in different technological domains, the usage experience for a specific service should be as uniform as possible across individual access technologies. Note that this does not preclude making use of advantageous features of a particular technology, such as larger bandwidth. We shall discuss this further in Part Two.

Service quality should be predictable also across different access technologies.

The ability to enjoy predictable service quality is increasing in importance. In order to gain true end-user acceptance, availability and quality of services and delivery methods should minimize variation during service usage session on one hand and between different sessions on the other. In this area, it is necessary to pay attention to both the absolute service quality level and the predictability of service quality (Räisänen, 2003a). We shall return to this issue in Chapter 6.

Authentication of different access domains and services provides a usability challenge for the end-user. Currently, the end-user must memorise a number of passwords, user identifiers, and Personal Identification Numbers (PINs).

1.2.2 Trends

We shall next review selected related industry trends related to packet-based services. More details about activities within particular industry and research fora will be given in Chapter 3. We shall use the same classification as for state-of-the-art description for structuring the trends.

Technology

The need to improve the agility of operations has been identified as a consequence of tough competition among service providers. To throw in another buzzword from the 2004 TeleManagement World in Nice, cost pressures are also leading to lean processes. Terms of the trade, these two fashionable concepts equal to requirements of flexibility and keeping operations resources at the optimal level in view of the task at hand. The overall target is the ability to identify what is viable from the business viewpoint, and selecting efficient design, implementation, and operation mode for the next phase of the process. This enables not only efficient operation for present-day services, but also addressing smaller, previously non-feasible market niches in relation to both market segment size and temporal duration of services (Koivukoski and Räisänen, 2005).

The above-mentioned trend has led to the pursuit of unifying BSS and OSS. Linking technical systems more closely to business management is an obvious enabler for streamlining operations. Achieving such a linkage requires gaining more common ground conceptually, procedurally, and from the information modelling viewpoint, as we shall see later on. It is necessary to study what is needed in practice to get rid of out-of-date, artificial boundaries.

Market development has led to the need for improvement in efficiency and flexibility of operations.

An example of this is the emergence of service management of services as a discipline in its own right. Managing services as separate entities helps in bridging the gap between BSS and OSS (Figure 1.5). Naturally, efficient operation of service management requires support from BSS and OSS. Focusing on services, and starting with end-user services in

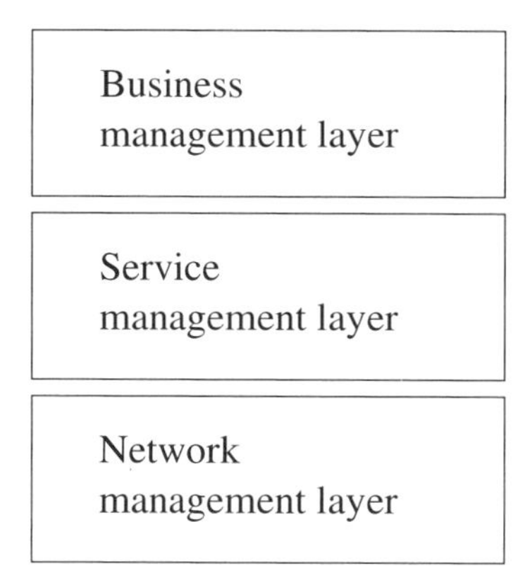

Figure 1.5 An illustration of the positioning of service management

particular, provides practical guidance for the paradigm change in the form of a set of down-to-earth requirements.

Over the years, it has become apparent that service management needs to be understood within the process context (Koivukoski and Räisänen, 2005). As we shall see later, understanding the service management phases and the relation of service management users is a valuable tool in this. An example of this will be provided in Chapter 3 in connection with TeleManagement Forum (TMF) service framework.

From the information management viewpoint, a trend towards shared representations of information is visible at the moment. In recent years, the adoption of various formats based on eXtensible Mark-up Language (XML) has been a simple example of this. More advanced endeavours target harmonisation of information models, in addition to streamlining of the protocol basis. Common information models are of help in coping with multi-vendor challenges that typical providers face. They also help in storing information related to services in repositories, for which term service inventories could be coined in loose analogy with network inventories.

Parallel to pursuing common information models, aiming at standardised process modelling has also proven to be useful. As we shall see later on, this is one of the areas in which industry has shown increasing interest. Broadly speaking, useful results in this area have been achieved by aiming at recording Best Current Practices (BCP) descriptions. Such an approach can be used, and indeed has been used in many different areas by IETF and TMF.

The importance of industrial cooperation is increasing.

The technology used for implementing and providing packet-based services is undergoing a series of paradigm changes. The technological environments used for providing services are being standardised – witness the degree of interest towards service enabler related work within Open Mobile Alliance (OMA). Of course, the same reason lies behind the success of GSM telephony and 3GPP frameworks in a broader sense.

The standardisation of platforms on which services are being operated is also expected to lead to the optimisation of operator's service creation processes and easier implementation of new services. We shall return to this later on. The service platform standardisation is parallelled by activities striving towards modularised designs in networking technologies such as Open Base Station Architecture Initiative (OBSAI).

A more fundamental conceptual change, there has recently been much interest towards technologies that are amenable to distributing functionality both within an organisation and across stakeholder boundaries. Design-wise, the currently favoured approach to this is based on decoupling of business logic from implementation of technical components. This means that the components need to provide clear interfaces towards the outside world, but the logic with which the components are used should not be dictated by the implementations of the components themselves. One of the names under which this work is being done is Service-oriented Architectures (SoA). Related work is also being done under the title of Service-oriented Computing (SoC) (Papazoglou and Georgapoulos, 2003).

Keeping business logic and component implementation separate from each other helps in integrating providers' systems more closely with each other. This is a potentially powerful enabler for organisational agility, but requires a protocol basis to match the requirements of the paradigm. Web Services Interfaces (WSI) provides one such technology. The hope is that WSI technologies can be used for building of completely new ways of interacting in addition to systems integration tasks.

Service-oriented architectures hold the promise of integrating systems and businesses in novel and innovative ways.

Systems supporting true convergence across multiple Internet access technologies are coming to the marketplace. Within 3GPP, adding support for WLAN access domains is part of the Release 6 content. Parallel to this, support for multiple access technologies is being addressed in IMS.

Broadcasting technologies which allow distribution of content to a large number of communications endpoints in an economical way are being piloted. Digital Video Broadcasting for Handhelds (DVB-H) is an example of this. DVB-H allows for distribution of mass content in a way that is economic for both the end-user and the content provider.

Web services are starting to also make good of their promise from the point of view of end-user services. At the time of writing this (May 2005), it is reported that a new service has been created by combining Internet real estate services with map services. Both of the services that existed separately in the past did not have specific connections to the new service. The new service allows the user to move about a city map and seek the locations of available apartments. This is but a glimpse of what is possible with web services.

Even a short review like this one would not be complete at the present time without a reference to peer-to-peer (P2P) services. Endpoints such as Personal Computers (PCs) at home or mobile devices can run programs like web servers. End-user services such as VoIP can be implemented with a direct connection between the endpoints involved. The accessibility of services run at endpoints can be enhanced by directories and support for registering to directories, as is the case with certain Internet telephony services. The inherent requirements of services set some boundaries on the quality of peer-to-peer services.

Peer-to-peer services need to accounted for in future service management systems.

Operation of networks is increasingly provided as a service. These kinds of services are provided by many mobile and IT domain vendors today, and also by specialised service providers.

Business models

Partly a consequence of cost and timetable pressures, partly facilitated by emerging technologies, there is also a clear trend towards increasing use of 'co-opetition', i.e., cooperation in a competitive setting (Koivukoski and Räisänen, 2005). Standardisation bodies have been a long-standing example of this. In addition, there have been many examples of competitors in one field cooperating in order to complement standards on specific technology issues during the recent years.

There is a clear need to deepen integration between different business entities. In mobile networks, both service providers and access providers would benefit from the ability to create new chargeable services quickly and in an automated fashion. Also, facilitation of real-time information exchange concerning service usage would clearly be useful. From the business viewpoint, closer integration of business processes in this respect requires support for automated modification of technical parameters without having to carry out renegotiation of the business agreement. To achieve this, traditional agreements are expected to be complemented with frameworks for electronic business.

Deeper business integration between business entities is a current trend.

The increasing use of IP as a unifying protocol platform for services on the one hand and the increasing automation of business relations on the other makes it possible for new kinds of 'lightweight' service providers to compete with incumbent parties in the areas of both existing and novel services. The various Internet telephony providers are probably the most famous example of this. At the simplest, only registration on a website and a broadband connection is needed for making international voice calls. The situation is not to be interpreted as being downright worrying for the incumbent players, however. The ability to support session-specific service quality parameters in an access domain may turn out to play an important role on a longer run. Session-specific support for service quality in Figure 1.4 is an example of this.

End-user

The systems currently being taken into use and being designed provide progressively better support for hiding the details of access technologies from the end-user. The methods supporting mapping of services to equivalent service quality in different technology domains are powerful enablers of this. Another example of hiding unnecessary technological details is the adoption of single sign-on technologies, relieving the end-user of the need to register to different access domains and to different service providers separately.

It is probably the case that not all the characteristics of the access domains are hidden from the end-user. It may even be desirable that end-users continue to appreciate basic differences between access domains, such as available throughput. In the interest of usability, it is desirable that the set of memorised new concepts is as small as possible, and preferably representable using familiar metaphors. As an example of the latter, during a recent trip in Central Europe, the author mentally drew a parallel to arriving at an oasis for reaching 3G coverage area for the cellphone.

The ability to specify and automatically apply personal preferences is expected to be of increasing importance. These kinds of technologies build on and generalise the already existing building blocks such as the ability to convert content to match terminal capabilities. Such a functionality is most likely an essential ingredient in maintaining ease of use for powerful new context-based services.

End-user viewpoint: irrelevant access technology-specific details should be hidden from view while retaining visibility to relevant ones. Management and applying of personal preferences should be supported.

Personal information management is growing in importance. The ability to store metadata related to content such as pictures and video recordings allows for the use of powerful search methods for personal content.

To provide longer perspective to service management, we shall next take a look at some emerging technologies that are being studied at the moment.

1.3 Emerging Technologies

In the previous section, changes in the characteristics of the access channel were indicated as being an important factor for wide-scale adoption of packet-based services. There is no reason to believe that this basic situation will change in the future. However, we are probably not too far off the mark in saying that emphasis will be shifted towards the way access technologies are integrated into a business environment.

The importance of the cost of the spectrum was referred to above. Operating services over licensed spectrum band affects the way systems are built, so that costly resources can be used in the best possible manner. Studies have indicated that radio spectrum is not used efficiently at the moment, whereby Federal Communications Commission (FCC) in the U.S. as well as other organisations have been active in initiating activities to enhance spectral efficiency. The dynamic sharing and trading of spectral resources, and their relation to Software Defined Radios (SDR) are being studied e.g. within the European Union (EU) Sixth Framework Programme (FP6) WINNER project (Winner, 2006). These kinds of developments, should they become reality, can be expected to affect the price of licensed spectrum on one hand, and make the distinction between licensed and non-licensed spectral resources less clear-cut.

The application of automated market-based mechanism in access technologies has been studied for some time now (Kelly, 2000), and different mechanisms for utilising them have been studied e.g. within the context of Differentiated Services (DiffServ) framework (Kilkki, 1999; Ruutu and Kilkki, 1998). In addition to applying market-based mechanisms at networking technology level for packet prioritisation and back-pressure mechanisms, full-blown systems for brokering network access have been described in the literature (Cortese *et al.*, 2003; Semret *et al.*, 2000). These kinds of technologies have the potential to facilitate more efficient utilisation of network resources. The application of market-based mechanisms has been studied for communications endpoints, in addition to various providers involved. The communications endpoint technologies for negotiating access to the Internet have been described in (Personal router whitepaper, 2006). These kinds of paradigms have the potential to make the relationship between end-users and providers more dynamic.

The growing complexity of services, as well as the parallel diversification of value nets involved in the providing of services, have led to a number of related activities for working out the consequences for the end-user view to services on one hand, and for related provider parties on the other. The overall goal is to provide increased automation of service usage and provisioning, while enabling a richer variety of services. An example of a type of a service with increasing importance is community participation (Churchill *et al.*, 2004).

The WSI framework referred to in the previous section provides building blocks for composing and using services dynamically. However, in the present form the core is based on a registry-oriented scheme and does not provide for intelligence in composing services in a variable environment. An example of WSI put into a more general framework can be found in (Ferguson *et al.*, 2004). A broad class of concepts for adding intelligence to services is called Semantic Web (Berners-Lee *et al.*, 2001; Davies *et al.*, 2004), where the goal is to develop a system where intelligent agents can compose services on behalf of the users.

Emerging trends include software-defined radios, market-based mechanisms in the network, and automated service provisioning systems.

We shall next describe two research activities outlining elements of the future of packet-based services.

1.3.1 WWRF

The Wireless World Research Forum (WWRF) is a global forum for seeking longer-term consensus on wireless technologies on Beyond 3G (B3G) era and disseminating the results of the work. Industry and academic institutions both participate in it. It has provided guidance for initiating research programmes. The ongoing EU projects within the Wireless World Initiative (WWI) programme are examples of this.

The WWRF has a working group for service architectures, analysing the technical basis for providing a platform for distributed service management. The latest results of the work have been recorded in (Tafazolli, 2004). In addition to the architecture working group, WWRF also has working groups for analysing user-oriented and business aspects of future services, cooperative and ad hoc networks , new air interfaces, short-range communication systems, and reconfigurability. Results of WWRF work are available in (de Marca *et al.*, 2004).

An interesting aspect about WWRF is that the entire work programme is built on the principle of user-centricity. The WWRF analysis of user-centricity starts with a basic

Maslow-style hierarchy of subsistence, safety and security, self-actualisation, and human capability augmentation. The user values relevant to WWRF are feedback, consistency, control, and privacy. The values listed need to be balanced against convenience of use. The value-add capabilities of systems supporting include natural interaction, context-awareness, personalisation, as well as ubiquitous communication and information access.

1.3.2 MobiLife

Another example of ongoing project-level activity, the EU FP6 project MobiLife is studying near-future packet-based services from the end-user viewpoint (Aftelak *et al.*, 2004; MobiLife, 2006). It is part of the WWI programme referred to earlier. The work within MobiLife is organised into work packages (WPs), which include user experience WP, personal area WP, group context WP, wide-area WP, and an architectural WP.

The utilisation of context information in wide-area, group-oriented, and personal sphere constitute broad use case classes in MobiLife. Relating to the provider view to services, the MobiLife project also has an architecture working group, which has studied the management of services in the MobiLife era. Relating to the role of peer-to-peer services, the MobiLife architecture allows for linking services by combining constituents from 'carrier class' operators and end points.

In addition to the study of the actual service architecture, MobiLife is also addressing some procedural aspects of service management. Figure 1.6 shows the first draft of a generalised service life cycle. A topic we shall encounter later in the book, life cycle provides an overview of the types of activities encountered during the lifetime of an individual service. The MobiLife service life cycle complements the telecom-oriented management view by facilitating WSI-oriented dynamic composition and usage as part of the generalised life cycle.

A spin-off of MobiLife architectural work, the relation of service life cycle to evolving service management paradigms has been analysed in more depth in (Räisänen *et al.*, 2005). The paradigms identified in the article include:

- 'Stovepipe' management. This is the present-day situation where one type of service is managed separately from the other
- Component-based service management. This phase constitutes a step forward by basing management of components which can be integrated with processes
- Distributed service management. The precise meaning of this phase is still subject to research, but it is expected that run-time composition of services will be characteristic of this phase.

The entire life cycle of services should be considered in designing service management.

Analogously to WWRF, the MobiLife project also addresses business aspects. This includes the identification of stakeholders to service provisioning, as well as the analysis

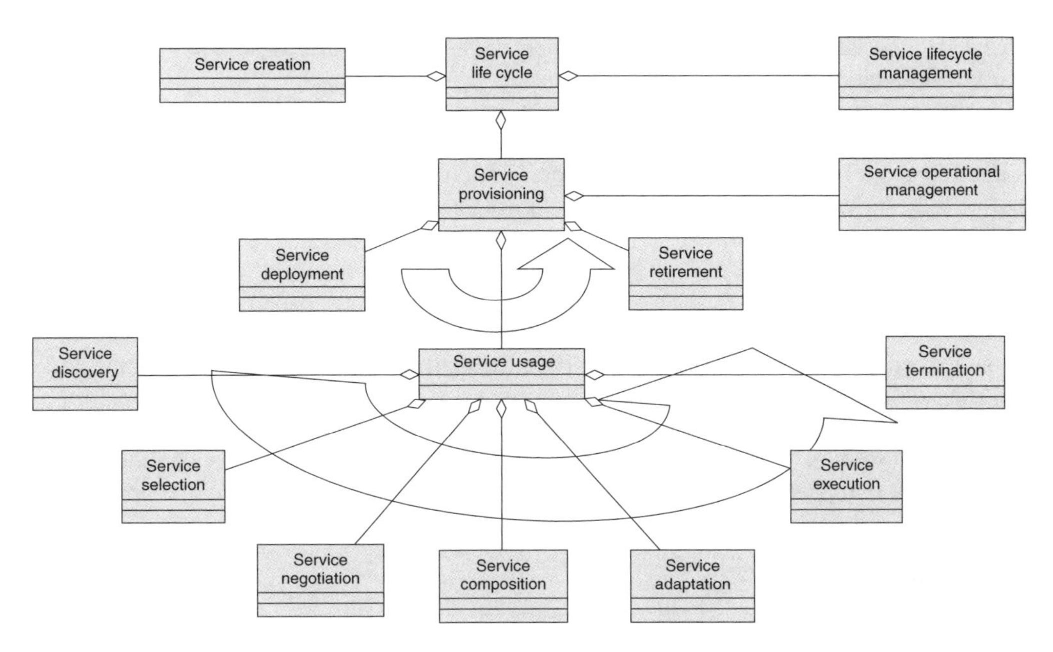

Figure 1.6 MobiLife service life cycle. Reproduced by permission of the Mobilife Consortium

of evolution of business models of individual stakeholders. The MobiLife project also contributes to WWI program wide functions, such as operability hurdles analysis.

1.4 Summary

In this chapter, we have provided a first definition of service modelling and discussed justification for the need of treating it as a separate discipline. It is an overview of relevant state of the art and trends within the industry.

The challenge lies in managing simultaneously the effect of the increase in the complexity of services, the increase in the complexity of networking technologies, and the anticipated diversification of business value nets. All this needs to be carried out in an environment where timeliness and efficient execution are important. The future-proof service management needs to cope with conflicting requirements, such as using streamlined processes on the one hand and being able to manage increasingly complex services in increasingly diverse technical environments on the other.

A clear trend resulting from the drivers listed above is the shift from a paradigm where the creation of a new service requires new hardware and software towards the service platform type systems where the end-user services can be created by managing parameters. IMS, together with capabilities in 3GPP core networks, provides one such platform. The generic service platforms are studied in a more general setting within OMA. Should the expectations associated with this kind of work be met with, they would not only simplify service management but also modelling.

The parallel activities ongoing in WWRF and EU FP6 addressing the reconfigurability (E2R, 2006) and composition (Ambient, 2006) of networks also have an impact on service management. Interfacing of service management to the access network should become easier because of these kinds of developments. These kinds of developments also affect the business models of the players involved.

Many new technologies potentially affecting service management are visible in the horizon. The good news is that different forms of international cooperation have been started at an early phase, well before standardisation, to assess potential future challenges before they surface. The cross-project operability hurdles analysis of the WWI programme is a good example of this.

Up to now, we have been describing services and service management of packet-based services from a relatively generic viewpoint. In the following chapter, we shall enter the domain of modelling with a summary of relevant paradigms. In Chapter 3, we shall study some related industry initiatives. In Part Two, we shall describe our modelling framework, building on the groundwork built in Phase one.

Perhaps the biggest challenge in devising a model relates to selecting the most adequate presentation format. The discussion in Part One sets the stage for the service model to be presented in Part Two, and made more concrete using the examples in Part Three.

1.5 Highlights

Ten things to remember from this Chapter:

- Service modelling is needed for streamlining service management.
- Service modelling amounts to modelling inter-relations between products, services, and resources.
- Service modelling needs to cover information entities and processes.
- The stakeholders participating to value net need to be taken into account.
- The goal of generic service modelling should be independent of specific business models.
- The existing and emerging service platforms such as IMS and service provider environments developed within OMA make it easier to create and operate services, but also pose challenges.
- The end-user perspective is important in service management.
- Striving to unify OSS and BSS is perceived important by the industry.
- Distributed service paradigms bring new possibilities and challenges.
- The work ongoing within EU FP6 MobiLife project and WWRF programme study business models, frameworks, and technology relating to future service platforms.

2

Approaches to Modelling

In the previous chapter, we provided an initial definition of service modelling without saying how modelling is performed. In this chapter, we shall study paradigms which help in putting the modelling framework to be presented in Part Two into a wider perspective. In view of the preceding text, this chapter provides a modelling-centric view which has a wider area of applications than service modelling. The issues discussed in the preceding and current chapters will converge in the next chapter as well as in Part Two.

2.1 Introduction to Modelling

Modelling – without further qualifiers – is a broad discipline. The author of this book is feeling humble as a consequence of the sheer magnitude of the challenge of trying to present a summary of modelling related to the topic of the book, let alone attempting to define modelling in a broader context. The basic reason for this is that the same problem domain can be modelled in very different ways. In what follows, we survey some of the relevant tools at the modeller's disposal.

To quote the Oxford English Dictionary (OED, 1995) again, a model is

> a simplified description of a system etc., to assist calculations and prediction.

Analogously, modelling amounts to

> devising a model of a system.

Judging by the above definitions, a model captures a useful part of the features of a system, whereas other parts not relevant for the purpose at hand are not modelled. Modelling should thus start with the definition of the system in question, and the definition of a model should include an account of the features included in the model on one hand and those that are left out on the other. Philosophically speaking, the description of the system can never be complete, but is meant to be interpreted here as providing a list of characteristics of the system which is larger than the scope of the model. The criterion for deciding which features to include results from the envisioned purposes of use of the model. Very good models can also be used for purposes not originally designed for, but

Service Modelling: Principles and Applications
Vilho Räisänen

one cannot hope to achieve this in all cases. In 'real life', usefulness often carries more weight than theoretical merit.

Models can be used for different purposes, including system design, conceptual verification, system verification, and as a basis for information exchange. The role of models in software development has been likened to that of blueprints in construction (UML, 2003). The 'blueprints' approach can be contrasted with 'theories of everything', the difference between the two being that blueprints intentionally stop at a certain level of detail, whereas theories of everything go deep into details. To provide an example of detailed modelling in engineering, packet-level simulation models can be devised to emulate the impact of a particular network loading situation on service quality experienced by a mobile end-user, for example. In this book, we are mostly interested in 'blueprints', rather than simulated models.

A model is a simplification of reality for a specific purpose.

Generally speaking, desirable properties for a model are simplicity, formality and intuitiveness. For simplicity, Albert Einstein's famous quote provides a guideline here: *Mache Dinge so einfach wie möglich, aber nich einfacher* (Make things as simple as possible, but no simpler). Obviously, the complexity of the model is affected by the system in question, as well as potentially the purpose of the use of the model. The degree of formality depends on the anticipated use of the model, and it is not desirable in all cases to put formality in a position of top priority. For example, for the purposes of rendering a model for user interface, amenability to human ways of working may be more important than formality, an oblique reference to which fact has been made within the motto of this part of the book. Multiple models of the same system can be employed to satisfy the needs of different purposes of use. Finally, the model should not have built-in redundancy, but should conform to Occam's razor principle to the best possible extent.

Cognitive aspects of modelling have been found important in research literature. Indeed, often, amenability to human perception is more important than accurate reflection of the real ontological structure of the world (Parsons, 1996). Note that applicability of this principle is not limited to user interfaces, but these aspects also need to be considered in other situations in which humans need to study the model in question. In physics, the behaviour of a harmonic oscillator (exemplified by a weight hanging from a spring) is understood very well, and therefore, surprisingly, many problems in physics are modelled as having some sort of harmonic oscillator in it. Making use of what comes naturally is not a bad idea, as long as it serves the purpose at hand.

One set of conclusions from educational experiments has been compressed in the form of a list of 'C's (Mayer, 1989):

- Completeness
- Conciseness
- Coherence

- Concreteness
- Conceptuality
- Correctness
- Consideration.

Especially for complex systems, it is often useful to devise a model consisting of multiple views. Correct choice of views can reduce the complexity of the model. As we shall see below, frequently used views include information organisation, behavioural view, and temporal view.

A model for a particular system can be devised in different ways. A particular version of a model is often called a *schema* (from Greek $\sigma\chi\eta\mu\alpha$, meaning a figure). For devising 'blueprint' kind of models, it is often desirable to describe the modelling of a system on an abstract level, so that different modelling alternatives can be compared. This can be achieved with meta-modelling, which models the features of the schemata, or different instances of a system model. One way of viewing the role of meta-modelling is to say that the meta-model provides the rules and grammar for the schemata (called *modelling languages* in this context) (Henderson-Sellers, 2003). In this paradigm, schemata are viewed as being instances of the meta-model, and relations between the entities within the meta-model are paralleled within ones in the schemata.

A model can be layered, with the highest-level model potentially instantiated in multiple low-level schemata.

We shall next take a look at some modelling paradigms. The sortiment of paradigms may seem baffling at first, but we shall summarise their relevance to our topic at the end of this chapter. To provide a quick idea, software development as a process is not completely unlike service development, and experiences thence are also useful for us. Object-oriented modelling is a prime example of this, and in addition considers expert systems, service-oriented architectures, database design, and architecture design methodology, among other things.

2.2 Software Engineering Paradigm

Software can be developed for different purposes and in different ways. The 'traditional' approach to software development is concerned with the development of a new piece of software or a software suite, which is sold to multiple customers as a stand-alone entity. Another type of software is developed for a particular environment as a tailored solution. A further variant of software is open-source software, which is developed by a distributed community. In what follows, we shall discuss each of the types briefly.

The traditional commercial software development paradigm is based on the identification of market needs and developing software to meet that need. The need in question maybe an anticipated future need or an existing one. The challenges with the traditional method lie in correctly identifying the market needs, defining the software to be developed, and allocating resources to the project in an optimal way. The patriarch of software development paradigms is the waterfall model, which more or less consists of the following phases:

1. Definition of business requirements
2. Definition of technical requirements
3. Technical design
4. Implementation
5. Testing
6. Deployment.

In real life, there is also an installation and after-sales phase which is not shown above.

Experience accumulated over multiple decades has shown weaknesses in the literal following of this paradigm. Perhaps the most important one is the observation that development of commercial software is seldom a purely linear process, but typically requires refinements to assumptions made in previous phases. Thus, in real-life software projects, there are often iterative elements at play. This trend has been highlighted with the growing complexity of software being developed. According to the Institute for Electrical and Electronics Engineers (IEEE) Spectrum, some versions of the Linux operating system contain 200 million lines of code.

The second important lesson is that not all phases are created equal. It is much cheaper to fix problems in the early phases rather than in the late phases of development. A third note relates to the management of business requirements. As the global marketplace seeks to increase its efficiency and the sheer volume of competition grows, the risk that business requirements change because of competition and changing customer needs also grows in importance and probability. An optimal software development paradigm must cater for this procedurally.

The biggest challenges with traditional software development paradigms are requirement specification and change management.

In tailored software projects, on the other hand, the environment is typically more stable. Customer requirements may still be subject to change, but their financial effects can be mitigated with liability clauses during the agreement definition phase. Otherwise, challenges in moving from requirements down to deployment are basically the same as in the traditional paradigm.

In open source projects with a large enough developer base, it is easier to achieve well-tested software than in closed, resource-starved projects. In some cases, the outcomes of the open-source projects have been truly excellent, of which GNU/Linux based systems are good examples. The challenge lies in attracting a critical mass of developers to the project as well as retaining participant loyalty. If an open source project relies on voluntary participation only, changes in commitment to the project may vary with time. Recently, other dangers have also surfaced; for example, it has been found that it may not be a good idea to trust that a particular tool central to the project remains available for free indefinitely.

Common to all of the three cases described above, support for software products is important. This is particularly important for open source software for which the normal after-sales organisation does not exist. Due to this, there is a market for companies which provide support to accompany open-source software, typically combined with a specific packaging of the open software. This model is popular with GNU/Linux distributions, for example.

To improve the waterfall paradigm, various proposals have been made over a long period of time. We shall not review these here, but mention two newer paradigms as examples, namely, Extreme Programming (XP) and User-centred Design (UCD). Typically, improvements to waterfall paradigm amount to coupling customer and end-user or other related stakeholders closer to the early phases of the development process. A specific challenge in involving customers earlier in the design phase, maintaining the stability and consistency of the overall solution needs to be paid attention to. A recent record of a less-than-successful story in this front can be found in (Goldstein, 2005).

Re-usability of software is important in all of the paradigms listed above. Standardisation of interfaces is an important enabler for integrating software from different vendors into the environment of the customer. The present trend appears to be away from tightly defined Application Programming Interfaces (APIs) and towards lightweight interfaces supplemented with semantics definition.

An important dimension of the software model is the relation between the supplier and the provider. According to (Brereton, 2004), the following phases in the development of the relation can be identified:

- Commercial off-the-shelf (COTS) software
- Component-based engineering (CBE)
- Software service engineering (SSE).

In the first phase, the customer's system is composed of commercially available standalone software products. In the second phase, software components are sold instead of complete programs. In the third phase, services are purchased instead of software from a global marketplace.

It is predicted that software development will increasingly move from COTS towards service engineering.

One consequence of the software engineering trends such as the ones described above is the fact that the importance of information modelling is increasing. It is not sufficient to just have standardised interfaces, but moving towards the world of networked services also requires that information exchanged through the interfaces in question can be managed flexibly. The same concerns are also relevant to modern network management systems, as discussed earlier (Martin-Flatin *et al.*, 2003).

It has also been argued (Jones, 2005) that facilitation of the information exchange between participants to the software development process plays an important role, especially when moving towards new, distributed service paradigms that are discussed below. 'Design by contract' type of thinking is viewed to be important by the author of the article as a device to avoid different developer parties from having conflicting views of how a service should work.

The 'blueprint' approach referred to above is especially relevant to software development. The following two requirements typical to this paradigm will be found to be of more general interest later on in this book:

- Implementation independence. The model should not be platform-specific, but should lend itself to implementation on any kind of computing system
- Traceability. The model should support linking to implementation and design phases of the software process.

2.3 Object-Oriented Modelling

Object-oriented Programming (OOP) has been popular since its inception as it supports reuse, separation of implementation from interfaces, and clear organisation of entities and methods assigned with them. Programming languages such as C++ provide for mechanisms for implementing basic aspects of OOP (Strostroup, 1997). The modelling part of OOP, Object-oriented Modelling (OOM), can be viewed as providing a modelling paradigm in its own right. Central to OOM is placing emphasis on the importance of entities and their inter-relations. Accumulated experience over the years has shown that certain patterns of modelling are useful. We shall discuss some of them in the following pages.

Perhaps the most widely agreed-upon aspects of object-oriented modelling include:

- Aggregation
- Generalisation
- Use of viewpoints.

Aggregation is a relation in which an entity aggregates a number of other entities. In the world of railways, an entity of type 'train' aggregates one or more engines and one or more railway cars. It is a special type of dependence between two entities.

Generalisation refers to a relation in which multiple entities can be aggregated into a single, more general entity. An OOP model for railways may only include Diesel engines at first, with electrical engines to be added at a later phase. In the latter of these two phases, the two types of engines can be generalised into an entity of type 'engine' which contains the features which are common to both subtypes. Having defined a generalised entity, the object-oriented modeller can proceed to derive further subtypes. This is usually called *inheritance*, and allows the 'inheriting' properties of the generalised entity.

The use of multiple viewpoints or views is fundamental to the modern use of OOM paradigms, especially in software development. Frequently, the following viewpoints are used:

- Static view
- Use case view
- Interaction view.

Static view describes the 'ontological' inter-relations of entities such as aggregation and generalisation. Use case view describes temporal interactions between entities. A use case view describes how different entities are used and by whom. Interaction view describes the order in which entities are referenced or invoked. The two latter views can be seen as providing a platform for 'table testing' and providing input to development of the static view, in addition to providing information about usage roles and temporal ordering. Different views refer to the same entities, whereby one is not talking about different models but perspectives of a single model.

Specific to software patterns, a number of commonly encountered OOP patterns have been identified (Gamma *et al.*, 2004). Generalising composite and atomic variants of an entity to an abstract entity is an example of this pattern. As an example of this, we could define abstract parent class for connected railway cars as `RailwayCar`, generalising `SingleCar` and `CarSection`, where the latter consists of multiple `RailwayCars`. We shall not describe them here in more detail, but make an advance note that parallel patterns can also be identified in 'pure' OOM which is applied to information modelling.

In OOP, a complete definition of objects typically includes attributes as well as the methods which can be applied to the object by external entities. They are part of the interface of an object, which facilitates hiding of the internal implementation of the entity and providing a clear definition of its interaction with other entities. For an imaginary entity of type 'engine', the method interface in C++ could look like this:

```
class engine {
 private:
  MaximumSpeed;
  ...
 public:
  AttachRailwayCarSection();
  DetachRailwayCarSection();
  GoForward();
  GoBackward();
  StopNormally();
  EmergencyStop();
  ...
};
```

The design of a system or information model using object-oriented paradigm is typically carried out in multiple steps, proceeding from semi-formal descriptions towards formal descriptions. As is the case of software designs, the object model can be usually improved based on experiences acquired in using an earlier version.

Object-oriented modelling employs relations such as aggregation and inheritance, and lends itself well to use of modelling patterns.

Object-oriented modelling is used by modern software development, since it provides value for large software projects. For example, it supports modular development by separating interfaces from implementation, and thus lends itself to be used for distributed development.

2.4 Expert Systems

Expert systems have their roots in Artificial Intelligence (AI), and seek to make solving of complex problems easier by providing more advanced ways of representing information than those available in conventional, general-purpose programming languages. When traditional programming means are used for solving problems, representation of complex problems often takes considerable amount of effort. Expert systems can help in solving problems within a limited domain with small 'encoding overhead'. To accomplish this, expert system typically needs to be specially built for a particular domain, which takes some implementation effort. One could liken the difference between expert systems and conventional programming languages to that of general-purpose data processors and Application-specific Integrated Circuit (ASIC). In both cases, lowering the generality of the environment makes it easier and faster to express procedures of a specific area.

Generally speaking, an expert system consists of user interface, knowledge base, inference engine, and working memory. Typically, expert systems are related to a specific domain of knowledge. The phases roughly analogous to the ones in software development can be identified for building an expert system. The types of roles needed for building an expert system include domain expert, knowledge engineer, and user. The domain expert provides the information that needs to be supported by an expert system. Knowledge engineer encodes the information in a format that can be used by an expert system. Feedback from a user is often useful in developing an expert system iteratively.

Rule-based programming is one of the paradigms that can be used in expert systems. In rule-based programming, rules are used for representing heuristics, being composed of triggers and actions. A trigger is a set of patterns which, when matched, trigger the rule. Actions, in turn, describes which actions are undertaken when the rule is invoked. In order to evaluate whether a rule needs to be triggered, facts must be compared with the trigger. This is called *pattern matching*.

Expert systems codify domain-specific knowledge to achieve highly expressive language for a specific set of tasks.

The relevance of expert systems to service modelling lies in the embodiment of the principle of aiding domain-specific tasks with tailored interfaces to the related information. While tasks related to service management can be carried out with general-purpose tools, paying attention to specific needs of service management provides significant value.

2.5 Service-Oriented Architectures

Service-oriented architecture (SoA) is a paradigm guiding the design of next-generation service architectures. It describes how systems participating to services are implemented, and the way that they interact with each other. There are many different definitions for SoA existing. In what follows, we try to capture some of the most essential features from the viewpoint of the present book rather than attempt to provide a complete academic description.

An essential ingredient of SoA is describing an architecture as a collection of relatively autonomous services. Individual services can communicate with each other via well-defined interfaces. Services can invoke other services, and higher-level services can be built by combining other services. Individual services are not necessarily visible to end-users. Services in the SoA sense can be used as building blocks for end-user services.

A direct consequence of the above description, SoA facilitates reuse and modularity. Another oft-quoted guiding principle for service-oriented architectures is decoupling of business logic from implementation of the components. This means that the interfaces provided by the components should be such that they provide maximum flexibility for carrying out business transactions. Such a system is viewed to provide a good basis for integrating business processes of individual stakeholders in an automated manner.

Conceptually, SoA can be said to consist of four layers (Hill, 2004):

- Presentation layer, managing interaction with end-users
- Process layer, executing business logic
- Service layer, consisting of reusable functions
- Systems layer, consisting of systems supporting services.

The service layer is often said to encapsulate underlying systems or to expose them as a web service.

Typical implementations of SoA include the means of discovering and using services as well as defining the ways in which the services interact with each other. The means of supporting communication between individual services are needed.

In SoA, functionalities are represented as services in ways that are suitable for use in different kinds of business processes.

From a theoretical viewpoint, the composition of services in terms of other services can be either static or dynamic. Static definition is easier to cope with in the early days of SoA adoption that we are living in at the moment. Subsequently, statically composed services are used in systems integrations by exposing individual subsystems as services in a Universal Description, Discovery, and Integration (UDDI) registry. Dynamic integration using registries is also possible in today's systems in a limited sense, and can be used for load sharing purposes, for example. The goal of employing 'fuzzy' descriptions of services and the ability to use non-exact matches takes us to the domain of Semantic Web, and is currently still a research topic (Lassila and Dixit, 2004). Related technologies are also being studied within the MobiLife project referred to earlier, for example.

Even though support systems for software development are not really within the scope of this book, the practically oriented reader may be interested in a recent article which – among other things – conveys the message that adhering to SoA requires support from development tools (Jones, 2005). The same source also stresses the importance of standardisation and common information models for SoA.

Finally, it has been pointed out (Foster, 2005) that some kind of a cultural change is needed to accompany the technical aspects of SoA to make the best of the service-oriented paradigm. This reflects the more general trends related to changing the way services are provided, referred to at the beginning of this book.

2.6 Databases

Database design is an established discipline in the field of information modelling and as such, provides a useful background for the undertaking that this book relates to.

Database design consists of data objects, the relationship between them, and operations that can be applied to them. Database design can be said to consist of the data model and the functional model. The former defines what is stored in the database and in what format, whereas the latter defines how stored data can be processed.

The dominant thinking in databases is the relational database paradigm, the history of which extends back to 1970. A relational database is composed of two-dimensional relational tables, the columns of which have pre-defined meanings. A row of an individual relational table constitutes a record, and is identified uniquely by a key field within the realm of a table. Keys can be used for linking information in multiple tables.

Relational tables are the format in which information is presented to the person or system accessing the information. The internal storage format of data may be different. When information is stored in multiple data structures, certain aspects of importance emerge. In the following paragraphs, we shall discuss integrity and normalisation.

Integrity means that it should be possible to access data stored in a database consistently, irrespective of the organisation of the data. The management of the keys of the record plays an important role here, since they are used for identifying records and relations between relational tables.

Normalisation refers to a process which seeks to remove the redundant information in relational tables and ensure that they can be modified consistently. It is also said that normalisation removes anomalies in the database structure. Different degrees of normalisation have been identified in research literature, and are called *normal forms* (NFs). Roughly speaking, higher-order NFs dispose of increasingly indirect anomalies. Five NFs have been identified, but a database does not necessarily have to be normalised to the fifth normal form (5NF). The most often referred-to normal form is the third one (3NF), which facilitates operations such as adding attributes to a record which does not yet have associated entries in other relational tables.

Relational modelling as a paradigm is not fundamentally orthogonal to object modelling, but the combination of the two has led to the perception of the existence of class challenges which have been called 'object/relational impedance mismatch', an oblique reference to the field of audio engineering. The extent to which this is only a mindset problem is still subject to discussion at the time of writing this. An example of the challenges involved relates to the optimal representation of the inheritance relation of OOM.

Matching relational models with object-oriented ones is not always straightforward.

2.7 Architecture Design

The need for architectural design of systems can be viewed from different angles, depending on the disposition of an individual. On the one hand, some Internet-related standardisation bodies such as Internet Engineering Task Force (IETF) have explicitly announced that they are not going to standardise architectures. Viewed superficially, the SoA paradigm would also seem to lead in this direction.

On the other hand, increasingly complex software systems clearly need an architecture to facilitate design and implementation. Also, systems based on IETF protocols need to be based on an understanding of the role of individual protocol interfaces. In other cases such as 3GPP systems, architectural work has a central role. If modelling corresponds to the blueprints of a building, systems architecture shows its relation to other buildings and facilitates the design of infrastructure such as water pipes, electricity, and so on. As such, architecture is also a model, even if a high-level one.

The Institute for Electrical and Electronics Engineers (IEEE) has produced a brief but interesting recommended practice document for architectural description of software-intensive systems IEEE Architecture Recommendation (2000). It encompasses conceptual framework and architectural description practices. The recommendation is meant to be used in creating architectures.

The conceptual framework has aspects such as definition of the purpose of the architecture, definition of the system, listing of involved stakeholders, and their viewpoints and concerns. Architectural description practices, in turn, describe how the conceptual framework is put together for a particular architecture.

The IEEE document describes the process of devising an architecture for a software system. In the actual architecture design, adequate tools of the trade such as use cases, system specifications, interface specifications, and so on can be applied.

It is perceived that the process of formulating an architecture is important for achieving good results.

The increasing complexity of systems and services run on them has underlined the importance of studying the requirements of a system as a whole, not only the subsystem or interface levels. Architectural work brings cohesion to a large system, and makes it easier to align the development of subsystems which the overall system is constituted of.

2.8 Other Modelling Methods

Some major modelling disciplines and schools of thought having relevance to the topic area of this book have been reviewed above. There are certain specific modelling techniques or patterns that are also related. A few of them are discussed below.

The 'Law of Demeter' is a design rule which is used in designing object-oriented systems. The rule has derived its name from the Greek goddess of agriculture, presumably referring to the stepwise 'cultivation' of object models. The most oft-quoted form of the Law of Demeter is 'only talk to your friends', or in more complete form 'only talk to your friends sharing your concerns'. In the original form, the Law means that objects should only have knowledge of the part of the object model that is closely related to them. In terms of services, the interpretation could be that services should only have knowledge of the services they need for their operation. One can see a link to SoA thinking here.

Model/view/controller is a pattern which is frequently used in Graphical User Interfaces (GUIs), for example. What is needed in these kinds of applications is managing partial, possibly also multiple views on data used by the application. The 'model' part of the pattern maintains the data, the 'view' part renders a view of the data for a particular use, and the 'controller' handles events relating to models or views. This is an example of breaking 'top-level' services down into distributed functions.

The division of a system model into management layer, control layer, and user layer is frequently used, in communication systems, for example. The user layer is concerned with the user data path. The treatment of user data is affected by the control layer in real time. The behaviour of the system is defined using the management layer. The management layer is typically considered to function in off-line mode, even though it is possible that a new configuration is put into action immediately. Referring to the 3GPP examples earlier in this book, static service provisioning would be performed using the management layer, reflected in the control layer of the GPRS Gateway Support Node (GGSN) as configurations of APNs. APN configurations determine mapping of user layer traffic to Packet Data Protocol (PDP) context and treatment applied to traffic.

Communication systems are often represented in the form of management layer, control layer, and user layer.

2.9 Summary

In this chapter, we have covered some approaches to modelling. We started with software development, since this discipline is also based on modelling in the sense that software design is a model of the desired functionality. It turns out that many aspects and techniques of software development can be utilised in modelling, especially when complemented with the bigger picture of life cycle involved. We shall address this bigger picture in Section 5.3.

Relating to service management, process modelling is important, in addition to information modelling *per se*. Above, software development as such is more oriented towards processes, whereas other viewpoints discussed relate mostly to information modelling.

Within the software development paradigm, the waterfall model in its original form has been found wanting in the modern development environment and has been suggested to be replaced with iterative techniques. These kinds of changes give rise to novel pitfalls which one needs to be aware of. The same requirement is relevant to information modelling in general.

Object-oriented paradigm, constituting a subset of Entity/Relationship (ER) models, is also applicable to information modelling. The meta-model/entity/instance division turns out to be especially useful for information modelling in general. Object modelling allows the use of generic patterns, an observation that has provided guidance for the central theme of Part Two of this book.

The database paradigm is well versed in terms of redundancy and anomaly prevention. The mapping of object oriented models to database representation needs careful planning where relevant.

Expert systems is perhaps not a paradigm on its own from the modern viewpoint, even though it has contributed to the development of technologies and process models which are useful, perhaps even increasingly so. Comparing rule-based programming to principles of policy-based management (PBM), for example, reveals the same basic rule structure in both. What is different is the way that the rules are formulated – PBM relies on formal rule structures, whereas expert systems often allow for more 'fuzzy' descriptions. What continues to be immediately useful in expert systems is the orientation towards the user of the system and codification of knowledge concerning particular areas to support the user.

The SoA paradigm seeks to represent functionalities as services with well-defined interfaces towards outside form. It sets requirements on the representation of information relating to individual services. In addition, SoA facilitates reuse in a natural manner. The practical ways of working related to putting SoA into practice are partly still being sought at the moment.

The use of architectures and related working procedures is important not only in software design, but also in information modelling. The use of views and separation of concerns bring benefits to both designing complex software systems and complex service models.

Generally speaking, information models are often viewed as consisting of the following layers:

- Concept model
- Logical model
- Physical model.

The concept model defines the 'universe of discourse', which is then refined in a particular domain in the logical model. The physical model defines details pertaining to run-time data. Depending on the requirements of the process, further layers can be employed. The use of meta-modelling layers for the concept model is an example of this.

We would like to remind the reader that it is not the purpose of this book to tackle business modelling in great depth. Only the aspects relating to basic value net are described below. Subsequently, actual business models are not in scope here.

A note needs to be made about Product Data Modelling (PDM), also known as Product Life cycle Management (PLM). It has not been dedicated a separate section above, since the author does not consider it to be a separate modelling paradigm, but rather, using the common techniques of the trade and – in effect – constituting a subset of service modelling. Similarly, Enterprise Data Warehousing (EDW) is not described herein separately. In the same way, event-based modelling of business processes (Hollander *et al.*, 2000) can be viewed conceptually to be in the same generic class as PBM, since it defines triggers and actions.

All of the modelling approaches summarised above have commonalities. They all pertain to modelling, but approach it from different viewpoints and thus place emphasis on different issues. From the viewpoint of service management, the processes related to software development are not fundamentally different from the ones related to creation of packet-based services. To address the entire field of service management, however, one should compare service management to the entire life cycle of software products and the processes relating to managing an IT infrastructure. The analogy is further enhanced by the trend towards SSE referred to above.

Some differences between the software development paradigm and service management can be perceived, however. One of them is the increasing focus of service management on end-user segment value. This also affects the management of related technical configurations. Software development, on the other hand, has typically been geared towards providing value on a 'strategic' rather than a 'tactical' level and is consequently associated with longer timescales than service management. These differences may be minimised, should paradigms such as XP be embraced by the software industry.

Another difference results from the fact that service management increasingly relates to the management of configurations instead of the building of new systems for new services. As discussed above, this transition is facilitated by the employment of advanced service platforms in service provision. This means that the responsibility of creating the service is shifting from the software developer toward the party responsible for managing systems, be it the provider's own staff or an external party.

The bottom line is that it is a good idea for service management to tap into accumulated expertise in the area of software development and other disciplines where they add value. As we shall see, for example, the use of multiple views brings value to service modelling.

In the next chapter, we shall take a look at some related industry activities and initiatives.

2.10 Highlights

Ten things to remember from this chapter:

- When models are used directly by humans, cognitive issues should be taken into account.
- Software engineering paradigm has aspects which are analogous to the ones related to service management.
- Object-oriented modelling can be applied both to software development and services.
- Employing of meta-modelling helps the structuring of information according to the degree of generality.
- Expert systems are an example of building a dedicated view for target domain.
- Service-oriented Architectures is a promising paradigm candidate for enhancing re-usability of service components.
- Anomaly avoidance is an important issue in database design.
- IEEE has created architecture design guidelines describing a process for creating software systems.
- In a distributed system, it may be useful to limit services' knowledge of each other to better structure management.
- Distributed service functionalities can handle specialised tasks, such as rendering the views of a model.

3

Industry Initiatives

In the previous chapters, we have reviewed the state of the art and trends in the industry as well as modelling techniques that have a relation to service modelling. We shall now tie these two trends together by describing industry initiatives which seek to utilise modelling techniques to address technological and business challenges. The descriptions in this chapter contribute to the requirements listed in Part Two, and subsequently are not limited to service modelling only.

3.1 Introduction

We shall begin with a few words about the positioning of industry initiatives with respect to standardisation and the needs of the industry.

The importance of standards have long been recognised within IT and mobile industries. Even the commercial players who have previously been interested only in proprietary implementations or variants of existing standards as *de facto* standards, have recently shown interest towards adhering to cooperation in the form of standardisation. Furthermore, the importance of open standards – which are not based on exclusive membership – has been increasing. The success of standards of Internet Engineering Task Force (IETF) such as Transfer Control Protocol (TCP) and Hypertext Transfer Protocol (HTTP) is a testament to this.

The reasons for this development have been described in Chapter 1. Parallel reasons are also contributing towards an increase in the interest for cooperation in a wider context than standardisation. There are a number of industry cooperation fora which seek to generalise and extend the work done within individual standardisation organisations (SDOs). These kinds of activities complement the liaisoning activity which is bread and butter for major SDOs.

Examples of new kinds of activities include generalised architectural work, which spans a number of access technologies. Activities of this kind have addressed the utilisation of non-licensed spectrum for accessing mobile services (Unlicensed Mobile Access, (UMA, 2006)) and the study of Beyond 3G (B3G) technologies (WWRF, 2006). Among other similar activities, Wireless World Research Forum (WWRF) and EU R&D projects belonging to the Wireless World Initiative (WWI) programme provide platforms for the general study of technology prior to actual standardisation taking place.

A class of activities is also addressing the methods used in describing the systems of the future, in addition to providing system designs. For example, Object Management Group (OMG) and TeleManagement Forum (TMF) are involved in this kind of work. Even though the basic set of available technologies is well known, achieving technical solutions, which are robust from the viewpoint of changing business environments, benefits from the participation of a large number of experts with different viewpoints and backgrounds. The risk of building complex systems out of subsystems that are difficult to integrate with each other is thus mitigated, at least in part.

Industry cooperation fora working in the areas of methodologies and information models have seen an increase in the level of interest.

In what follows, we shall take a look at the work done within individual fora and summarise the findings at the end of this chapter. We shall take a look at methodology fora first, and then move towards more technology-specific ones.

The following questions are important in view of formulating a service model:

- How is information modelling represented and with what tools?
- How are processes represented?
- How are packet-based services modelled?
- In what way can a multi-access network be represented?
- How are the different phases of the life cycle of a service represented?

The above questions provide a way of keeping track of the positioning of different activities with respect to the topic of this book while reading the descriptions. We shall summarise our findings at the end of this chapter.

3.2 OMG

OMG (2006) is an open membership, not-for-profit organisation, the *raison d'être* of which is producing specifications for interoperable enterprise applications. There is a special focus on distributed applications. The best known of these are Model Driven Architecture (MDA) (Miller and Mukerji, 2003) and Unified Modelling Language (UML) (UML, 2003). In addition, OMG has produced a specification for meta-modelling in the form of Meta-modelling Object Facility (MOF) (MOF, 2000) and an object management architecture. Object interchange format has been defined in the form of XML Metadata Interchange (XMI 2.0, 2002). OMG has also defined Interface Description Language (IDL, 2001) to accompany its middleware architecture, the Common Object Request Broker Architecture (CORBA) (CORBA, 2004). Mappings to IDL have been defined for languages such as C, C++, and Python. In the following pages, we shall start with the concept of meta-modelling to bind the different working areas together.

Table 3.1 OMG modelling layers. Adapted from (Henderson-Sellers, 2003) and (MOF, 2000)

Modelling level	OMG concepts	Example
Meta-meta-model	Meta-Object Facility	hard-wired meta-model
Meta-model	UML	description of structured data types
Model	User concepts	structured variable
Data	User run-time data	specific value for variable

3.2.1 Meta-modelling Viewpoint

The relation of OMG to meta-modelling has been shown to consist of four layers (Table 3.1). The relationship between the different layers relates to a 'loose' interpretation of meta-modelling (Henderson-Sellers, 2003). A reader interested in the strict interpretation of meta-modelling discoveries can consult this source.

In view of the earlier discussion about meta-modelling, the two middle layers of 3.1 are self-explanatory. The topmost layer is interesting, as it defines the language in which meta-models are devised. The lowest layer can be specific to a run-time environment.

Model-driven Architecture, in turn, relates to the way meta-modelling layers are used as a part of an entire system. We shall return to this topic later.

3.2.2 MOF

MOF is a meta-modelling language, the role of which can be viewed from two viewpoints: the definition of meta-models for a domain on the one hand, and the management of information within a particular domain on the other. The actual MOF specification, defining a set of CORBA IDL interfaces provided by OMG, describes a core set of a meta-model, which can be extended for particular uses. MOF can be used as a meta-meta-model for defining meta-models. Automating of metadata management in enterprise data warehousing is also listed as a possible use for MOF.

The current version of MOF has some restrictions compared to UML. Examples of these are support for binary associations only and the absence of association classes.

3.2.3 MDA

MDA was originally conceived to address the different aspects of the life cycle of applications, including design, deployment, integration, and management (Miller and Mukerji, 2003). The design goals of MDA include portability, interoperability, domain-specificity, and productivity. As we have seen before and will also see later in this book, striving to simultaneously satisfy the seemingly orthogonal requirements of generality on the one hand and domain-specificity on the other helps in bringing about added value.

The basic concepts of MDA include system, model, architecture, platform, and viewpoint. The model is a description or specification of the system for a particular purpose. The architecture consists of the specification of the parts, connectors, and interaction rules within a system. The platform refers to a set of subsystems providing a coherent set of functionality. The viewpoint is used as an abstraction technique to address particular concerns. The viewpoints of MDA include:

- Computation independent viewpoint, focusing on the environment and the requirements of a system
- Platform independent viewpoint, focusing on the operation of a system
- Platform specific viewpoint, focusing on the specific needs of a particular platform.

In MDA, types of models are devised to address the needs of particular viewpoints, including the Computation Independent Model (CIM), the Platform Independent Model (PIM), and the Platform Specific Model (PSM), corresponding to the three views listed above. Transformation is a process in which a particular type of model is converted into another. Meta-models such as MOF can be also be used for PIM and PSM, in addition to CIM, by applying adequate mappings. Mappings are also used to convert a PIM to PSM. Techniques such as the marking of entities and the application of patterns can be used in transformations.

MDA employs computation-independent, platform-independent and platform-specific viewpoints in modelling.

The above tri-partition seeks to address the conflicting requirements of generality and domain-specificity by moving from the requirements phase via a domain-specific viewpoint to a platform-specific viewpoint.

3.2.4 UML

UML is a semi-formal description language, which is suitable for meta-modelling and modelling purposes. It is important to understand the meaning of the expression 'semi-formal' in the context of UML. The different representation formats of UML provide a degree of formality as compared to non-structured representations, but are not alone sufficient in providing a rigorous model. There are multiple reasons for this, including the fact that it is not typically possible to represent the entire model in a single UML diagram. Multiple diagrams are needed to provide the views of the system in question, naturally leading to the question of the best way to choose the views used. Another reason is that the names of the entities and their relations in the UML diagram are not sufficient for fully documenting their meaning. Thus, typically, prose and examples are used to complement the UML model.

The UML specification (UML, 2003) lists business models and non-software systems as potential application areas, in addition to the most obvious use in software systems design. In addition to modelling-oriented goals, raising of abstraction level and integration of best practices are listed as the targets of UML. The specification expressly says that UML is intentionally process independent, and goes on to say that processes are typically enterprise-specific. We shall return to the last issue later in this chapter. UML

Table 3.2 UML 1.5 diagram type hierarchy

Diagram type	Subtype	Sub-subtype
Use case diagram		
Class diagram		
Behaviour diagrams:		
	Statechart diagram	
	Activity diagram	
	Interaction diagrams:	
		Sequence diagram
		Collaboration diagram
Implementation diagrams:		
	Component diagram	
	Deployment diagram	

documentation also stresses the role of visualisation in modelling. This has obvious links to our discussion about the cognitive aspects in the previous chapter.

To support the generic goals of modelling discussed earlier in this chapter, UML provides a hierarchy of diagrams listed in Table 3.2.

With reference to the four modelling layers discussed earlier, (Table 3.1), UML itself is described on the logical meta-model layer. As such, it addresses declarative semantics, to which relevant implementations should conform. The meta-model description of UML consists of diagrams, natural language, and formal notations.

The authors make a reference to the theoretical problems associated with using a language to describe itself – presumably referring to Kurt Gödel's work on logical foundations of mathematics in the first half of the 20th century – but go on to say that the scheme works in practice according to their experience. Overall, a UML package is described with abstract syntax, rules governing what is well-formed, and semantics. For the second of these, Object Constraint Language (OCL) is included as a part of the UML specification.

UML is a widely-used tool within the modelling trade, and good tools are available for various UML modelling tasks. Most UML tools allow for the generation of code 'stubs' from the UML models in multiple object-oriented programming languages such as C++ and Java(TM). As a semi-formal method, it needs to be put into a wider process context to produce rigorous results. Nevertheless, used adequately, UML provides a relatively well-defined context for discussing models and as such has proven to be very useful within its domain of applicability.

3.2.5 CORBA

CORBA describes a distributed architecture, in which object of a collection communicate with each other. The objects provide a way of supporting request-based invocation of services through an interface, hiding the implementation details from the interface for requesting services, constituting an implementation of object orientation principles. The messages are conveyed between the client and the server by an Object Request Broker (ORB). The interface to the relevant object is defined using OMG IDL. Communication between multiple ORBs is also possible. In addition to messaging, CORBA also includes functions related to the life cycle of objects.

CORBA has been included in this summary for two reasons: it is a product of OMG, and it is a representative of a class of distributed architectures. From the second viewpoint, CORBA is not the only alternative, Java™ Remote Method Invocation (RMI) is an example of other environments. Nevertheless, CORBA is a well-known reference the documentation for which is easily available. Distributed architectures are expected to have an increasingly important role in the future, as we have referred to earlier. In view of the interfacing paradigms, CORBA can be classified as being Application Programme Interface (API)-oriented.

CORBA describes an architecture for distributed computing.

Next, we shall describe an industry forum that has been devised in view of the needs of a specific branch of industry.

3.3 Business Process Management

3.3.1 Workflow Management Coalition

Workflow management is a process control discipline, amounting to a system for the breaking down of a task into subtasks, allocating these to personnel, and supervising the progress of subtasks.

The Workflow management coalition (WfMC) states that its mission is to promote and develop the use of workflow by producing standards relating to terminology, interoperability, and connectivity across individual products WfMC (2006). WfMC has produced a workflow reference model to serve as a basis for its work WfMC Reference Model (1995), Hollingsworth (1995). The reference model includes a description of components and related terminology.

Within WfMC, a workflow is defined as being a facilitation or automation of a business process, in part or in entirety. A workflow management system, in turn, is a system which defines, manages, and executes workflows such that the order of execution is driven by a computer representation of workflow logic.

The most important parts of a workflow management system in the terminology of WfMC reference model are workflow definition tool, enactment service, and applications. The first of these relates to the definition and allocation of subtasks, the second to the execution of these, and the latter to the interfacing to the workflow management system by other tools. Worklists are used for mediating workflow items between the enactment service and the participants to the workflow.

WfMC has provided standard definitions and reference architecture for workflow management systems.

WfMC provides a specific example of the benefits an industry forum can bring with it: common concepts and terminology can be devised, contributing positively to the bottom line of both the customers and the vendors. The benefits of this approach are highlighted by the growing complexity of systems and the business environment.

3.3.2 OASIS

Organisation for the Advancement of Structured Information Systems (OASIS) (OASIS, 2006) is a forum for advancing the development and adoption of e-business standards. It has produced a specification for web services discovery protocol called Universal Description, Discovery, and Integration (UDDI) protocol. OASIS has also produced specifications related to the way that web services are used. Business Process Execution Language (BPEL) is an example of this (Davies *et al.*, 2004). It provides the means of defining business processes in terms of coordinated web services. Recently, OASIS has also started work on the reference model on service oriented architecture (OASIS-SoA, 2005).

BPEL is not the only method for web services choreography, but it provides a good example of building added value on top of the Simple Object Access Protocol (SOAP)/UDDI/Web Services Description Language (WSDL) stack. The service model must accommodate these kinds of operations in one form or another. The addressing of the needs of an e-business is also an important viewing angle for service modelling.

3.3.3 BPMI.org

Business Process Management Initiative (BPMI) cites the need to build better and new processes faster as the first item in its list of drivers. Other drivers include an enhanced understanding of processes, and the automation of simple processes. To address these goals, BPMI seeks to define Business Process Management Notation (BPMN) and Business Process Semantic Model (BPSM) to build on W3C and OASIS components such as BPEL.

3.3.4 RosettaNet

RosettaNet cites the development of universal standards for the global supply chain as its mission. The reductions in cycle time and inventory costs as well as improved productivity are cited as practical expected outcomes.

The current list of RosettaNet standards relate to the matching of business processes between trading partners, related terminology, and implementation framework.

3.4 ITU

The International Telecommunications Union (ITU) is an international organisation for the coordination of global telecommunications networks and services, which governments and the private sector participate in. ITU produces international recommendations to be ratified by national or regional SDOs. The most well-known activity of ITU is probably the international allocation of licensed radio frequencies for technologies such as Global System for Mobile Communication (GSM) and Universal Mobile Telephony System (UMTS). The ITU telecommunication standardisation sector (ITU-T) produces recommendations for different areas of telecommunications.

ITU-T has produced a number of conceptual models, which are of interest to us. One of them is the modelling of communications bearer, which can be used to transfer information between Service Access Points (SAPs). As a consequence of the influence of ITU in traditional telephony systems, the original definition of bearer was related to connection-oriented bearers. Subsequently, the definition has been generalised to include connectionless bearers as well. (ITU-T Recommendation G.809, 2003). Connectionless bearers can be used for modelling routed networks such as Internet Protocol (IP) domains. ITU-T has also provided an analysis of the concept of Quality of Service (QoS), decomposing it to different viewpoints relating to the provider and the customer of the service (ITU-T Recommendation G.1000, 2001). We shall return to this definition later on in this book. There is also a recommendation listing multimedia service quality categories (ITU-T Recommendation G.1010, 2001). ITU is also working in other areas such as network management and next-generation multi-access communication systems.

ITU-T has produced conceptual models relating service quality and communications bearers.

From the viewpoint of modelling, ITU-T has done conceptual analysis about the interrelations of basic concepts, and thus contributes to the use of these concepts in different kinds of models. The deliverables of ITU-T have been disseminated widely, which is a definite advantage.

3.5 3GPP

The 3GPP is a collaboration agreement between Asian and European standardisation bodies. The scope of work includes the development of Wideband Code Division Multiple Access (WCDMA) and GSM-based networks into third generation systems. 3GPP not only performs extensive architectural and standardisation work for mobile systems, but has also extended its domain towards the integration of Wireless local area network (WLAN) and other non-cellular access technologies.

Directly relevant to the topic area of this book, 3GPP has a service provisioning model and QoS architecture which we shall summarise below. Please recall that we discussed the basic 3GPP service provisioning modes in the first Chapter. We shall also make some other references to 3GPP work later in the book.

The QoS architecture of 3GPP (3GPP TS 23.107, 2004; 3GPP TS 23.207, 2004) is based on the concept of an end-to-end bearer, which is requested by the communications endpoint. The attributes are provided as part of the activation request. The network may downgrade the bearer service quality level requested by the terminal. This results in a service quality negotiation procedure between the terminal and the network (Koodli and Puuskari, 2001; Räisänen, 2004). In the 3GPP model, the endpoint control of the bearer properties is a central ingredient. Through the control of the bearer creation and modification, the endpoint is also in control of the costs associated with connectivity.

The maximum service quality level allocated for a particular kind of a service can be provisioned to the network based on the type of a service in question as well as on the attributes stored in the QoS profile per subscriber. The use of the QoS profile relates to both the static and dynamic provisioning modes described in the first chapter. A service can be detected based on the IP classification criteria such as the IP address or a protocol number (Räisänen, 2003a). In 3GPP Release 5, the version supporting IP Multimedia Subsystem (IMS), the dynamic linking of session properties to the attributes of a bearer (Laiho and Acker, 2005; Poikselkä *et al.*, 2004) are supported. This can be called *service quality support instantiation* as a part of service instantiation (Räisänen, 2003a).

The 3GPP service quality framework is interesting for the reason that it provides true multi-service support and is also deployed at the moment. A number of service provisioning modes can be used with it, which makes it flexible. At the same time, it also presents an operability challenge in the form of a technologically advanced, novel, and relatively complex system in the present-day multi-vendor world (Koivukoski and Räisänen, 2005; Räisänen *et al.*, 2005). More information about the 3GPP QoS framework and service provisioning can be found in Appendix A.

In Release 6 (R6), 3GPP architecture is generalised to be less cellular technology specific. One example of this kind of work is Generic Authentication Architecture (GAA). Parallel developments can be observed in the activities targeting the interfacing of IMS to non-cellular access networks.

3GPP has produced a service quality framework encompassing both static and dynamic provisioning, and is working on convergence support.

3GPP brings valuable input to modelling both in terms of challenges as well as solutions. Regarding the first aspect, 3GPP – especially the third generation variant thereof – is one of the first widely deployed multi-service platforms. This brings with it a set of definite needs, which helps to make requirements for modelling more concrete. Since the deployment of 3GPP systems is at an advanced stage, solutions and systems are also available.

3.6 TeleManagement Forum

TMF is an international organisation which makes the improvement of management and the operation of information and communications services its goal. TMF seeks to provide a forum for formulating, recording, and disseminating Best Current Practice (BCP) descriptions, to be applied and extended according to enterprise-specific needs.

As discussed in Chapter 1, the streamlining of the concepts of BSS and Open Source System (OSS) has been one of the important work areas for TMF. The work has been carried out as a part of the New Generation Operations Support Systems (NGOSS) programme, in which representatives from telecommunications service providers, vendors, and system integrators participate. In addition, there are application teams that provide guidance for the NGOSS framework in specific technical areas. The main work areas of NGOSS include business process map, information modelling, integration framework, and conformance criteria. NGOSS considers systems development from four viewpoints: business, design, implementation, and deployment. The core NGOSS teams are supported by a number of application teams that provide input and direction to their work.

TMF formulates BCPs descriptions to be used as a basis in enterprise-specific environments.

TMF has addressed a number of topics over the years, and an example of their earlier achievements is the Wireless Service Measurement Team (WSMT) which addressed Key Quality Indicators (KQIs) and Key Performance Indicators (KPIs) for wireless systems (Wireless Service Measurements Handbook, 2004). Service Level Agreement (SLA) handbook (SLA Management Handbook, 2001) has been another working area. TMF has also produced a description of Technology-Neutral Architecture (TNA), based on distributed architecture providing implementation opacity via the use of components and interfaces (Technology-Neutral Architecture, 2004). The separation of business logic from implementation has been stated as one of the goals of TNA. A methodology for moving system designs from business level definition to system design, to implementation, and on to deployment has also been described in (NGOSS Methodology, 2004).

We shall describe below some additional teams of the TMF that have relevance to our topic.

3.6.1 Enhanced Telecom Operations Map

The Enhanced Telecom Operations Map (eTOM) team is perhaps the most well known part of the NGOSS programme. It is a continuation of an older Telecom Operations Map activity, and has produced a business process framework (eTOM, 2004) that has subsequently been embraced by ITU-T. The goal is to describe a framework for all the business processes that a service provider uses. This also includes the processes related to inter-provider operations. In addition to process classification, eTOM includes process flow examples, and an appendix of eTOM describes mapping to IT Infrastructure Library (ITIL). We shall summarise ITIL work and its relation to eTOM later on in this chapter.

The eTOM work is organised into three main parts, or process areas:

- Strategy, Infrastructure and Product
- Operations
- Enterprise Management.

The first of the areas is responsible for creating saleable products and services, the second for their operation, and the third for the business management of the enterprise. Within each of the main parts, the processes are organised according to the process area as well as functionality. eTOM provides multiple views into the enterprise processes, such that the next level provides a refinement of the view of the previous level. The views are numbered so that L0 view corresponds roughly to the tri-partition described above, and L1-L3 provide progressive refinements to L0.

The eTOM model is widely used as a reference in telecommunications, and has been adopted by ITU-T.

Many companies have publicly announced that they use eTOM in their operations. Examples of these kinds of testimonies are available on the TMF website (TMF, 2006). eTOM is a good example of how process classification modelling can be carried out and used.

3.6.2 Shared Information/Data Model

Shared Information/Data (SID) model is an NGOSS programme activity related to information modelling (SID, 2004). SID builds on previous information modelling activities within IETF, Distributed Management Task Force (DMTF), Directory Enabled Networking (DEN), and ITU-T. SID can be mapped to eTOM, but does not currently cover the entire area of eTOM.

SID consists of an Aggregate Business Entity (ABE) framework and a number of addenda addressing specific types of information entities belonging to a particular business entity domain. Examples of top-level domains include market and sales, product, service and resource, analogously to the basic structure of eTOM. The detailed modelling of each domain is described in an addendum using a combination of UML diagrams and prose. The UML part of SID can be viewed to describe an ontology for related entities.

The concepts of product, service, and resource that are used in SID will be referred to later in this book, so a few words about them are in order. The product is something that can be purchased by a customer. Service is a technical implementation of a product, supported by resources. SID does not define a service in a rigorous manner, but states that it is tightly bound to products within SID. SID differentiates between Customer-Facing Services (CFS) and Resource-Facing Services (RFS), the former having aspects visible to the end-user and the latter not. Resources can either be logical or physical, the latter characterised as something that can be picked up. Logical resources, in turn, represent abstract functionalities such as operating systems or software.

The SID model develops a basis for information modelling, and supports representation of linkages between products, services, and resources.

SID can be mapped to Common Business Entities (CBEs) of the OSS Through Java™ (OSS/J) initiative, to the extent that relevant entities exist in SID (OSS/J White Paper, 2004). OSS/J CBEs may also constitute extensions to SID.

As a model, SID does not cover everything. However, SID is based on the experiences and output from the prior modelling efforts referred to above. This is a definite advantage of SID, and one of the main reasons why SID is widely used as a reference. SID provides detailed examples of how its framework can be applied to fixed networking technologies, including Virtual Private Networks (VPNs) and Multi-Protocol Label Switched (MPLS) networks. SID also gathers together a number of useful modelling patterns, which have been found useful within the industry. In addition to inter-relations of product, service and resource, aggregation and role patterns will be referred to later on in this book.

3.6.3 Service Framework

eTOM and SID address the description of processes and information relevant to the business of the service provider. As such, they do not specify how they could be used together to address practical problems. This constitutes a real challenge, given the breadth of the area of applicability. To get a better insight into these challenges, the Service Framework Team (SFT) was founded. It is one of the TMF application teams referred to above. It studied the TMF tools available from the viewpoint of service providers and came up with some concepts of its own. The relation of SFT to the work of some other TMF teams is illustrated in Figure 3.1.

As described in the previous chapter, some of the basic challenges in service management are associated with the order in which tasks are carried out, as well as with the related process linkage and information ownership issues. SFT has approached the issue via service management roles, analysing their relation to the service life cycle. Service creation as part of the service life cycle was selected for more detailed analysis. The detailed analysis relates to the order of the appearance of roles in the course of service creation and identification of relevant eTOM processes and corresponding SID aggregate

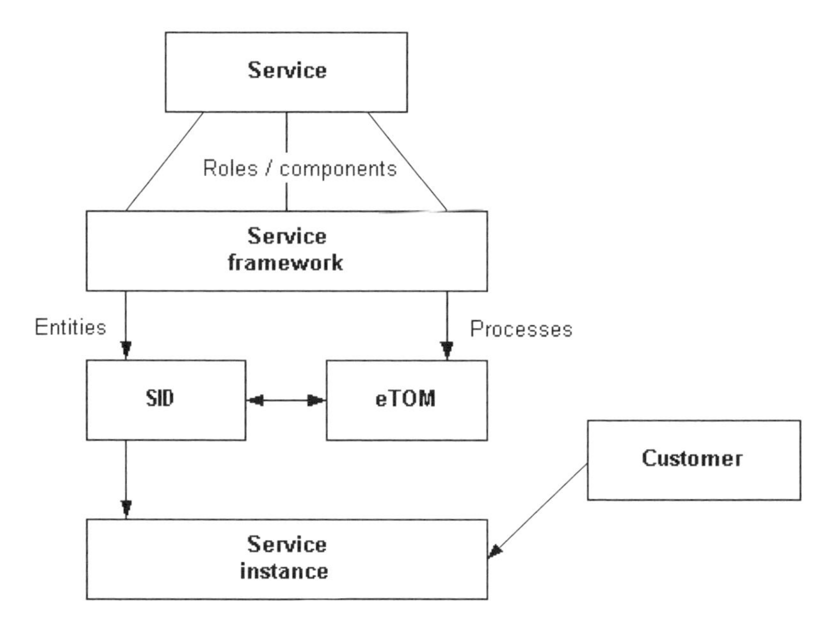

Figure 3.1 An illustration of the role of the service framework to SID and eTOM. From (Service Framework, 2004). (Reproduced by permission of the TeleManagement Forum)

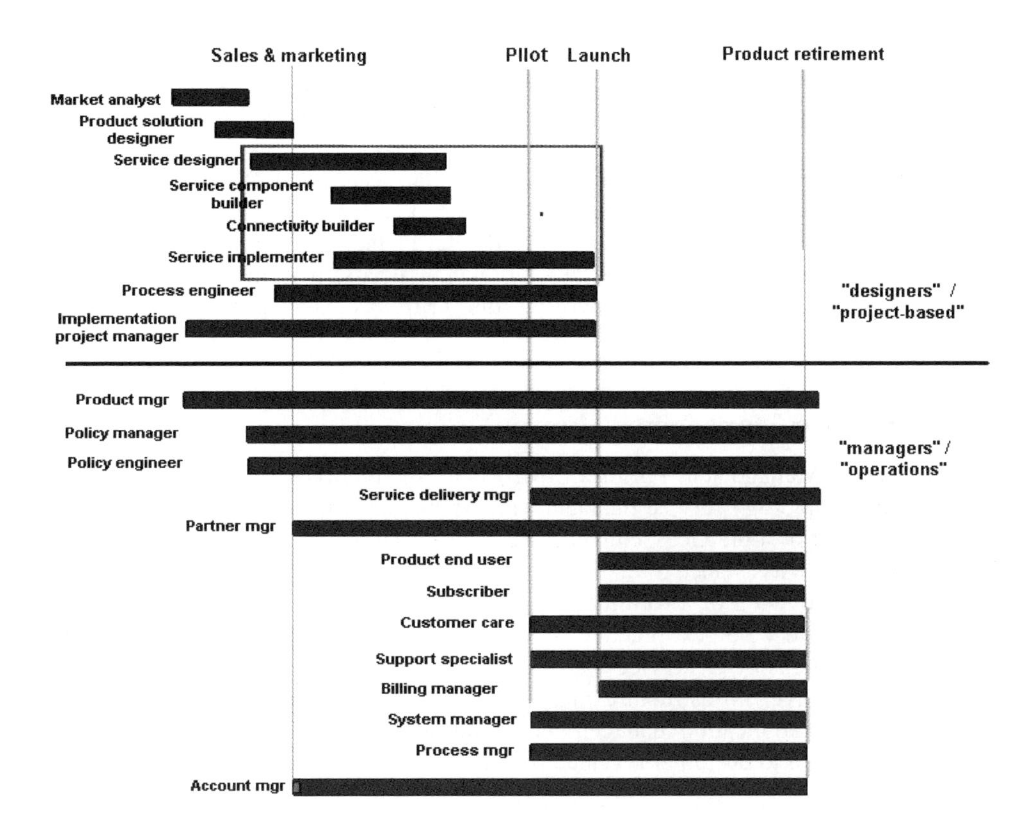

Figure 3.2 An illustration of the participation of service management roles to the creation and operation of a service. The red rectangle indicates the area of detailed analysis. From (Service Framework, 2004). (Reproduced by permission of the TeleManagement Forum)

business entities associated with the roles (Figure 3.2). A role in this context is really a set of activities, which can be carried out by one or more employees, and a single employee can participate in one or more roles.

The SFT team was founded to study practical requirements for the rest of the NGOSS work.

From the viewpoint of modelling *per se*, SFT has not really brought in new modelling patterns. What has been valuable, however, is the process via which the existing tools and techniques are employed in the course of addressing specific issues.

3.6.4 Services Over IP

Services over IP (SoIP) is another TMF application team, with the purpose of delivering a set of common requirements for the support of operational functions in managing resources and services (SoIP business requirements, 2005). The goal of SoIP work is to trigger the development and delivery of interfaces that can be used for procuring related systems. Rapid service provisioning in a multi-vendor environment, reliable service assurance, and tasks associated with Service Level Agreements (SLAs) have been listed as business foci of SoIP.

The importance of SoIP work to modelling is the description of patterns for services and resources in the document referred to above. The team has used Voice over IP as a first case study in their document. As with 3GPP, this first application area of SoIP work is associated with a definite set of practical needs in terms of the environment where it is deployed. At the time of writing this, TMF has extended SoIP work to address Next-Generation Networks (NGN).

3.7 IT Service Management Forum

The IT Service Management Forum (itSMF) is an international organisation working on Information Technology service management. As its goals, itSMF states the development and promotion of best practices, engendering of professionalism, the provision of a vehicle in order to provide assistance to its members for the improvement of service performance, as well as the provision of a forum for the exchange of information.

itSMF has produced a best practices document called *IT Infrastructure* Library (ITIL), covering broad areas of IT service management. These include implementation planning, business perspective, infrastructure management, service support and delivery, security management, and application management. An introduction to ITIL (ITIL Summary, 2004) states that attention to customer service provisioning is another core principle of ITIL.

itSMF has been devised from the viewpoint of IT infrastructure, and addresses a subset of eTOM processes.

TMF has provided a document describing eTOM/ITIL mapping (eTOM/ITIL Application Note, 2005). According to this document, ITIL constitutes a subset of eTOM, and has more IT oriented viewpoints than eTOM. Within its domain of applicability, ITIL is a frequently used reference. As to modelling, the IT paradigm should be taken into account for future systems in order to be able to best utilise their accumulated wisdom in that sector.

3.8 Activities Related to Internet Services

One of the most important area of activities relates to web services. ‘Web services’ is a vague term. Depending on the context of use, it can mean anything from simple web browsing to the use of semantic web technologies. An oft-quoted feature is the orientation towards simplicity and interoperability instead of detailed interface specification (ACM Queue, 2005).

Here, we are in the lucky position of being able to circumvent the potentially controversial definition of what web services are by taking a practical route. Multiple industry fora are working on specifications related to technologies which can be viewed as being related to web services. We shall briefly describe a few of them below, but would prefer to broaden the title to read 'Internet services' rather than 'web services'. OASIS also falls into this category, but we described it earlier.

3.8.1 W3C

Word Wide Web Consortium (W3C, 2006) produces standards related to web technologies. The most well-known standard produced by W3C is eXtensible Mark-Up Language (XML). XML Schema has been provided as a meta-level description of XML-based formats. For web services, W3C has produced the basic transaction specification called SOAP, as well as the specification for (WSDL). W3C has also produced the specifications related to semantic web, including Resource Description Framework (RDF) and Web Ontology Language (OWL). In addition, W3C has activities related to the usage of web technologies in a mobile environment, web services choreography and semantic web, for example.

The value of W3C for modelling relates to the tools and paradigms related to web services. Specifically, one can mention metadata description languages, semantic web, and technologies such as WSDL and SOAP complementing basic web services technologies. W3C standards are geared towards a lightweight approach to building distributed systems.

3.8.2 WS-I

Web Services Interoperability Organisation (WS-I, 2006) is an open organisation for the promotion of WS-I. It has produced specifications defining basic profiles for W3C and OASIS web services core standards (SOAP, WSDL, and UDDI), and profiling the standards for interoperability. WS-Policy provides an XML-based framework for specifying the capabilities and requirements of services independent of the domain. There are further specifications addressing trust and federation, for example (Mockford, 2004). In addition, WS-I also provides tools for testing profile implementations.

As an organisation within the field of web services, the role of WS-I can be compared to compliance activities within TMF and OSS/J. It provides a venue for combining the web services building blocks and identifying potentially missing parts.

Addressing of interoperability is important. The clarity about interoperability comes at a cost, though. Conformance to different aspects of the WS-I framework adds a lot of complexity to the simple basic web services framework of SOAP, WSDL, and UDDI.

3.8.3 Liberty Alliance

Liberty alliance (Liberty, 2006) develops open standards for privacy-friendly, secure and robust identity management. It has produced frameworks for identity federation, and identity web services, among other results. Liberty alliance has described an overall architecture, with service providers, identity providers, and users as basic building blocks.

Ever since the large-scale adoption of Internet Protocol-based services, the role of security seems to become more and more pronounced with time. As systems, services, and business value nets grow more advanced and distributed, the emphasis within the field of services also shifts. One needs to think beyond basic security technologies such as encryption.

Privacy and trust are issues within security, the importance of which is already clear now, for example, in the context of so-called 'phishing attacks'. Another driver for paying attention to the federation is increasing focus on the usability of services. The work of organisations such as Liberty are extremely valuable, and service models do well in supporting the stakeholders of the Liberty privacy and trust framework as well as the related methods such as federation.

3.9 Other Fora and Viewpoints

The organisations that have been listed above are perhaps the most important ones from the viewpoint of service modelling. However, related work has also been done in other fora. There are also some viewpoints which are interesting, but not related to a single industry cooperation. Below, we shall take a brief look at SLAs and OMA activities, and discuss the significance of semantic web services to the topic area of this book.

3.9.1 SLAs

Service Level Agreements (SLAs) provide the basis for agreeing on terms of delivery between the provider and the customer. Essentially, the SLA is a business contract, specifying service quality in technical terms, in addition to normal business contract related data such as period of applicability, definition of liability, and so on. The technical part of SLA, in turn, typically contains definitions of measurements of service quality relating to the service in question, as well as reporting procedures involved. The management of the terms associated with SLAs is important for business value net in current, increasingly dynamic environments. Subsequently, SLAs have also been studied elsewhere, in addition to the TMF working group mentioned previously.

The Differentiated Services (DiffServ) working group of IETF has proposed terminology related to SLAs in (Grossman, 2002). According to DiffServ WG terminology, SLA is the business agreement, and Service Level Specification (SLS) is the technical part of the agreement. In addition, Traffic Conditioning Agreement (TCA) and Traffic Conditioning Specification (TCS) are used to record assumptions about conditioning, which, roughly speaking, amounts to an agreement on the regulation of traffic variations between the parties to the agreement. The team has produced a further concept, namely, Per-Domain Behaviour (PDB), for specifying aggregate service quality levels within a DiffServ domain. We shall make use of these concepts in Parts Two and Three. Appendix B contains more information about DiffServ SLAs and PDBs.

IETF and TMF have done work in the area of SLAs.

Traditionally, SLAs have been used in the same way as other business agreements. The Creation and Deployment of END-User Services in Premium IP NEtworks (CADENUS) project of EU studied the creation and deployment of end-user services in IP networks with advanced, dynamic service quality support (Cortese *et al.*, 2003). The framework uses different kinds of mediator entities for automating service quality management, including access, service, and resource mediators. Access mediator interfaces to communication endpoints and service mediators within service provider domains. The interface between access mediator and service mediator can be viewed as the representation of dynamic SLA information. The information between the service mediator and the resource mediator, in turn, is more akin to the technical part of SLA, or SLS. The resource mediator is a bandwidth broker-type element, discussion about which can be found in (Räisänen, 2003a).

The significance of SLAs for service models is the ability to link service quality to the providers and the customers. The technical definition of SLA should lend itself to use for both business and technical use in the long run. It shall be possible to employ SLA-related concepts in the service model in the context of both traditional business context and automated, broker-type agents.

3.9.2 OMA

The Open Mobile Alliance (OMA) was founded in 2002, and multiple existing SDOs were merged into it. Mobile operators, wireless system vendors, companies working in the area of information technology, content providers, and others participate in it.

The mission of the OMA is the specification of mobile service enablers in order to enhance interoperability and ease of adoption of services. The goal of the OMA is to operate in an operating system and in an network technology independent manner. The results are documented in the form of open specifications. The OMA has technical working groups such as data synchronisation, device management, location technologies, presence, push-to-talk over cellular, and security. The OMA also has an architecture working group, which has described the OMA Service Environment (OSE). The work related to the service providers' platforms has been done under the OMA Service Provider Environments (OSPE), addressing a rudimentary set of service life cycle issues, in addition to the requirements relating to the component-based service platform as such.

OMA has addressed technology-independent service provider environment standardisation.

The OMA is a good example of the advantages of an open platform activity. From the viewpoint of modelling, OMA work brings with it the need to describe operations related to a standard service execution environment utilising components.

3.9.3 Semantic Description of Web Services

OWL-S (formerly Defence Advanced Research Projects Agency (DARPA) Agent Mark-up Language, DAML-S) is an ontology description language for web services which is based on RDF. The design goals of OWL-S include the facilitation of automatic web services discovery, invocation, composition, and execution monitoring. The top-level ontology of OWL-S consists of the service profile for discovery, the service model for the description of service operation, and the service grounding for describing the messaging to and from the service. The service model, in turn, includes the process ontology for describing how the service operates, and process control ontology for tracking the operation of a process. Further details can be found in (Davies *et al.*, 2004) or (DAML-S, 2003).

Other approaches towards the same goal as OWL-S include Web Services Modelling Framework (WSMF), which seeks to extend the capabilities of the OWL-S with formal semantics for describing state transitions for web services. It also places an emphasis on interoperability through mediation services. WSMF was developed by an EU project studying Semantic Web enabled web services. Currently OWL-S is a more widely embraced paradigm than WSMF.

Semantic web services technologies bring with them the need to model service descriptions which can be dynamically interpreted by agents, facilitating more flexible and powerful services than the present-day web services paradigm. Related technologies are studied, among other fora, in the MobiLife project referred to in Chapter 1.

3.10 Summary

It is time to summarise the results of our journey to the field of industry initiatives. Regarding the questions formulated at the end of Section 3.1, we can say that we are lucky in having a lot of input to service modelling from the activities reviewed. Good information models and patterns exist for both information and processes. The work of analysing packet-based services in terms of service modelling has begun, and the increasing importance of multi-access systems will doubtless bring with it more elaborate patterns in this area. Representations of life cycles have not been considered only for services, and also products and platforms used for provided services. This is all good news for our current undertaking.

We have reviewed the work done within a number of fora in which industry and academia have participated. Some of them are traditional, closed SDOs, whereas others are based on the concept of open membership. Increasingly, the importance of standardising information management schemes is recognised, in addition to standardising systems. In both cases, the work should be guided by an orderly architecture analysis. In addition to the standards, records of BCPs have been found to be useful reference points.

Fora such as OMG have provided better insight into what it takes to put SoA-type distributed architectures into action. While understanding of the basic mechanism is of great importance, it needs to be brought to a more detailed level for specific applications. Organisations working on web services related technologies, mobile systems, and operability focus have brought – and continue to bring – such detailed focus. Their work is complemented by research projects and programmes such as WWI, MobiLife, and WWRF that have been referred to earlier.

Run on top of IP, all service related transactions are composed of streams of packets. From this perspective, one expects that IT and mobile services would have a lot in common. After all, the IETF have 'marketed' IP as a convergence layer: 'everything over IP, IP over everything'. Indeed, technical studies have found many commonalities between different access technologies, a fact supported by activities such as UMA. On the other hand, paying attention to specific characteristics of access technologies helps in providing services in the most optimal manner at the present moment.

Similar dualism also exists in the business domain. On the one hand, service providers would like to offer services independent of the distribution channel. On the other hand, the history of IT and mobile domains, as well as the basic assumptions that the current systems are built on, are different from each other. Both the paradigms are in a state of flux at the moment, and one can expect that there will be increasingly more commonalities between the paradigms.

From the viewpoint of service management, at the moment, the IT and mobile domains are already largely sharing the goals of developing systems: information representation and process description should be standardised. Due to the differences in the way that IT and mobile services are provided to customers, the two paradigms are at different phases of development of unified concepts. Process maps have, nevertheless, found their way into both IT and mobile domains, and information modelling is also of increasing importance.

The principle of striving for generality as far as possible on the one hand and the requirement of providing PSMs on the other has been found to provide useful results. One needs to consider the way of moving from the general to the specific to bring best value to the user of the frameworks. The CIM/PIM/PSM structure of OMG and the four views of TMF NGOSS are fruits of these kinds of activities.

The activities described above have brought some insight into the best ways of using models in systems design. It is useful to use multiple models, addressing the specific needs of particular viewpoints and stakeholders. One must keep in mind that many of the models used are and can be informal, and have an important role in facilitating a more detailed analysis. The importance of visualisation in gaining acceptance for and enhancing understandability of models should not be underestimated.

The development of consistent and widely accepted models is paramount to increasing our understanding of what the best way of formulating a model both in terms of model structure, as well as its use in real-world environments is. In the course of modelling activities, the development and use of modelling patterns has proven to be very powerful. A good pattern can replace great amounts of work in many different specific models.

The current plethora of industry initiatives reflects the active development of new technologies and systems, facilitated in part by the Internet Protocol as the end-to-end platform for services. The many parallel activities have also highlighted the importance of coordinating activities in making the outputs of the various fora usable in practice. WS-I and TMF SFT are examples of these kinds of activities.

The emergence of new paradigms also sets new requirements for service modelling. The basic web services paradigm is based on simple messaging and the use of registries, which can be complemented with choreography definitions. The Web Services Interface (WSI) approach can be viewed to constitute a fundamentally different approach to building distributed systems as compared to API-oriented systems such as CORBA (ACM Queue, 2005). The distribution of services highlights the importance of privacy and trust technologies, as well as the models associated with them. In the long run, the ability to use partial information and combine it for added value brings with it great potential and also great challenges. Related technologies and models are studied within EU FP6 programmes, for example.

Parallel to the proliferation of technologies and paradigms, unifying tendencies can also be observed. Standardised service platforms are studied within the OMA, and there is a growing interest towards pre-standardisation activities covering multiple markets and technologies. WWRF is an example of the latter.

Overall, much work of interest and importance has been done, relating to service management in one way or another. In the following chapter, we shall try to formulate a framework which will bring the essential parts together.

3.11 Highlights

Ten things to remember from this chapter:

- In addition to UML specification, OMG has addressed meta-modelling and architecture modelling.
- CORBA is OMG's distributed architecture with objects representing services.
- WfMC has developed a reference model for workflow systems.
- ITU-T has developed modelling concepts for bearers.
- 3GPP has developed a multi-service QoS framework.
- TMF has addressed process modelling in eTOM and information modelling in SID as part of the NGOSS programme.
- SFT and SoIP teams provide input for the development of NGOSS from specific viewpoints.
- W3C, OASIS, WS-I, and Liberty alliance work in the area of web services.
- OMA addresses the development of modular service provider environments.
- Semantic web technologies are currently at the research stage, with their goal being the automation of web services.

Part II

Service Modelling Concepts

Scope of Part Two

In this Part, we shall describe a framework for service modelling, using the basis provided in Part one.

In the preceding pages, we discussed the needs of the industry, a number of available methods in the area of modelling, and related industry initiatives. This part builds on these descriptions, and seeks to select key issues in modelling among the material reviewed. Chosen viewpoint addresses some of the current needs of the industry, and outlines aspects of future modelling needs.

We shall start by summarising requirements for service modelling in Chapter 4, and continue by describing a management framework to be used for describing stakeholders' processes in Chapter 5. In Chapter 6, we shall describe a framework for describing technical aspects of services, and summarise requirements and characteristics of major classes of end-user services. This part will be concluded with a description of building blocks for a service model in terms of a framework and modelling patterns in Chapter 7.

The purpose of Part two is not to present a complete service model which would be suitable for all uses, but rather list viewpoints to take into account in constructing one. Chapter 4 lists generic requirements for a service model, and Chapters 5 and 6 provide tools for describing process-oriented and technical aspects of service models. Patterns described in Chapter 7 illustrate selected patterns in modelling.

We shall illustrate the use of service modelling concepts in Part three using examples.

4

Requirements for Service Modelling

In this chapter, we shall collect a set of requirements for service modelling. The summary mostly refers to the topics reviewed in Part One, but we shall also add some generalising and complementary elements.

We shall use multiple views for analysing and enumerating requirements. Individual views allow for detailed analyses, and use of multiple views brings perspective to requirement analysis. The requirements described below pertain to the service model as well as the way it is used. We shall call this overall area Service Modelling Framework (SMF). An example of addressing a subset of requirements in modelling is described in Chapter 7. The intention of listing the requirements here is not merely to be a lead-in to Chapter 7, but rather to enumerate issues to be taken into account in implementing a complete SMF.

The requirements described herein are high-level requirements, and are not meant to be detailed technical requirements. What is described below are the requirements towards SMF, unless otherwise stated.

Modelling of services can be viewed as having commonalities with modelling of software systems. The methods employed here include the use of stakeholders, viewpoints and concerns specific to stakeholders. Thus, the description of model does not follow the process of (IEEE Architecture Recommendation, 2000) in detail, but has adopted some concepts from there.

In the next section, we shall use the terms 'product', 'service', and 'resource' to refer to entities that are sold to customers, technical implementations required by products, and capabilities supporting services, respectively.

4.1 Notation

A brief note about graphical notation used in this part is in order. Justification for modelling is expressed in prose, and views of a model devised for a particular purpose are illustrated pictorially. We shall use simplified Unified Modelling Language (UML) notation for the latter purpose, the most important constructs used in this book of which

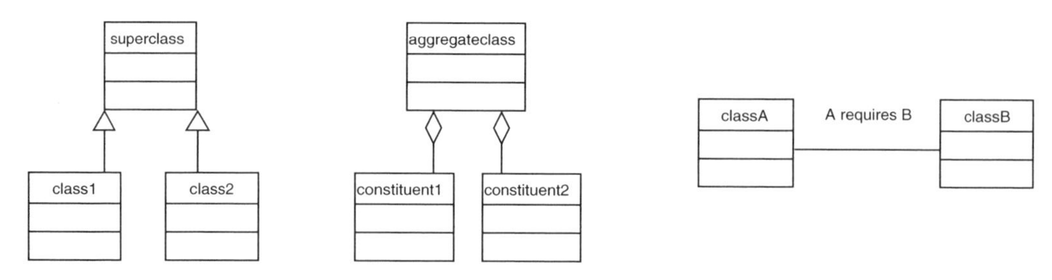

Figure 4.1 Static view UML notation examples: generalisation (left), and aggregation (middle), and dependence (right)

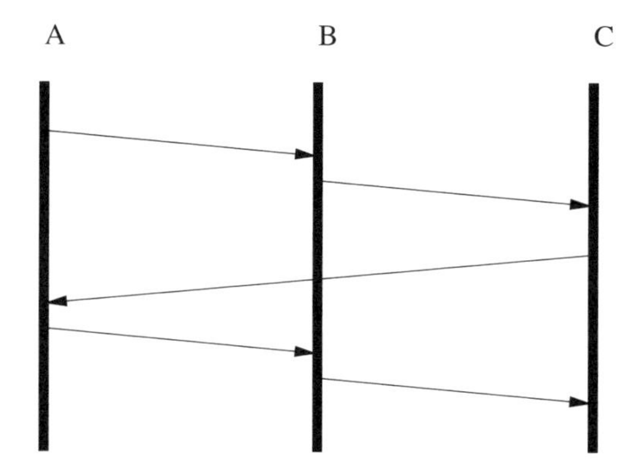

Figure 4.2 Dynamic view UML notation example: generic signalling sequence example involving entities A, B, and C. Time flows from top to bottom in the diagram

are shown in Figure 4.1 for static view and in Figure 4.2 for dynamic view. Views provide types of snapshots of the model as a whole.

The static view constructs used in our modelling diagrams allow for representation of generalisation, aggregation, and dependence between entities of the model. Generalisation is used for indicating that an entity is a special case of a more generic entity. Conversely, specific classes can be viewed as being inherited from more general ones. Aggregation shows explicitly that an entity is an aggregation of other entities. Dependence relation shows that there is another kind of inter-relation between two entities. Dependence relation can also be directed in UML, but we shall use only non-directed dependencies in our models.

The dynamic view constructs can refer to the same entities as the static view. The order in which resources are invoked during the dynamic view in question is indicated by a sequence of directional relations (arrows) between the entities. Parallel actions are allowed in the dynamic view. The dynamic view can be used for analysing the effect of individual phases of the sequence to overall flow of events.

In the examples of Part Three, we shall also utilise use case diagrams. The elements of use case notation are actors, use cases, and relations between them as shown in Figure 4.3.

UML has more advanced notations at its disposal, which are not used in this book in the interests of simplicity and understandability. For example, associations between

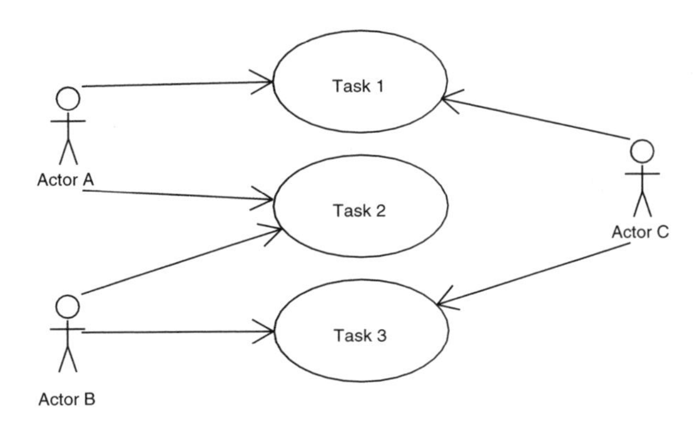

Figure 4.3 Example of use case, involving actors A, B, and C and their relations to tasks 1, 2, and 3

entities can also be classes. The quick summary of Figures 4.1 and 4.2 only captures the bare basics, and an interested reader is referred to (UML, 2003) or a suitable textbook on the topic.

We have also decided not to name relations and to leave out multiplicities at the end of relations. Attributes of entities are not described. In this book, we do not differentiate between containment and other kinds of aggregation.

The overall purpose of the notation is to outline relations of importance with a broad brush with the goal of covering a wide area, rather than to produce lengthy but detailed descriptions of a narrower area.

4.2 General Requirements and Concerns

We shall start by listing generic requirements. Related requirements are summarised in Table 4.1 and discussed in the following text.

An important type of generic requirements relates to the types of services supported. The framework should be future-proof according to the best current understanding yet

Table 4.1 Generic requirements and concerns

SMF should be constructed from the viewpoint of managing services.
SMF should not be limited to a particular technology or architecture.
SMF should not be limited to specific modelling paradigms.
SMF should support Service-oriented Architectures (SoAs).
SMF should support services provided with present-day architectures.
SMF should not be tied to any particular value network between stakeholders.
SMF should not be tied to a particular business model of individual stakeholders.
SMF should support linkage to service management–related processes.
SMF should support iterative improvement of the framework.
Formulation of SMF should not constrain the set of possible services.
SMF should support both operator-hosted and non-managed services.
SMF should support different phases of service management process, covering both high-level definitions and mapping to detailed, formal representations.

also provide support for present-day services. Thus, for example, framework should be sufficiently broad to cover future services such as the ones being studied within the MobiLife project. On the other hand, it should also be possible to apply the framework to present-day services such as instant messaging and chat, for example. The requirement of not restricting the set of possible services is of course a tall order, but best effort should be made to circumnavigate previously known hurdles. Paradigm-related requirements have also been listed, exemplified by a reference to SoA. The significance here in practice is that service building blocks should be amenable to definition of business logic separately from component implementation.

It should be possible to implement the framework using either future distributed service architectures such as those being studied within Wireless World Research Forum (WWRF) or on present-day architectures such as 3GPP service provisioning model and IP Multimedia Subsystem (IMS). The requirement of not limiting the applicable business models is important for facilitating the transition from single-provider services towards richer value nets.

Requirement of the ability to iteratively improve the framework relates to the fact that experience with prior conceptual frameworks suggests that it is seldom possible to produce the most usable model in one go. Therefore, it should be possible to elaborate the model on the basis of experience gained during the use of the model. Viewing the model as a tree branching from core concepts, such improvements would most likely relate to the 'leaves' of the model, rather than the 'trunk' parts. However, even larger modifications should not be ruled out, provided that there is sufficient justification for them.

The requirement to support non-managed services brings in a new dimension to service management, introducing services not provided by a traditional provider type stakeholder. Fashionable peer-to-peer services are an example of services belonging to this class. It should be noted that the inclusion of this requirement does not lessen the role of the traditional providers in any way – it would be fairer to say that it allows for a richer palette of possibilities for all stakeholders. Related further requirements will be described below.

4.3 Technology-related Requirements

The significance of technology-related requirements here is the ensuring of generality of the requirements with respect to technology platforms. Even though we have made plentiful references to 3GPP and IMS, for example, service management framework should work just as well in different kinds of technological environments, too. A few technology areas have been named, on the basis of the identified importance of Fixed–Mobile Convergence (FMC) and distribution, in our review of the state of the art and trends in Part One.

Technology viewpoint relates to systems that are used for accessing and providing services. From the technology viewpoint, the requirements listed in Table 4.2 can be identified.

The requirement of being able to apply the same model to both mobile and fixed network domains does not relate only to technologies but also to paradigms associated with the technologies. The reader is referred to the discussion on Enhanced Telecom Operations Map (eTOM) and IT Infrastructure Library (ITIL) in the previous part for an example.

Table 4.2 Technology-related requirements

High-level model should not be technology-specific.
It should be possible to map high-level model to technology-specific implementations.
The same model should support both mobile and fixed network domains.
It should be possible to operate the same services over multiple access technologies.
Tailoring of service according to access technologies should be supported.
Service quality control and management should be supported.
Security control and management should be supported.
SMF should not exclude simultaneous use of multiple access technologies.
Both static and dynamic service composition should be supported.
Both 'monolithic' and distributed service platforms should be supported.
It should be possible to mix managed and non-managed service components.

It is probably useful to point out that access independence and tailoring of a service to a particular access channel are not contradictory requirements. The former refers to the ability to provide a service via multiple access technologies, whereas the latter refers to making the best use of the capabilities of the access technology. For example, possible lack of support for guaranteed bit rates for video streaming in a Wireless LAN (WLAN) hotspot could be compensated with larger available bandwidth, access point load allowing.

The importance of the ability to apply the same framework to both management and control of service quality cannot be overstated. Service management refers to off-line actions required for managing services, whereas service control refers to on-line actions carried out by the network (Koivukoski and Räisänen, 2005). In view of both accumulated experiences and the goals of agility and leanness, easy mapping between management information structures and service control data is clearly an advantage. We shall discuss related requirements in Chapter 6.

Static provisioning refers to mechanisms such as traditional DiffServ Service Level Agreements (SLA), where service quality is statically allocated according to service type. Dynamic provisioning refers to the capability to dynamically allocate service quality for an entity of traffic in the form of instantiation of service quality support. We reviewed the two modes on a high level in Part One, and will discuss this further in Chapter 6.

Future service management framework should allow for composition of services by combining service components hosted by traditional provider stakeholders with components provided by stakeholders with non-managed environments. This allows for new kinds of roles for providers of managed services as supplying functionalities for peer-to-peer services, for example.

4.4 Process-related Requirements

Process-related requirements relate to the need towards modelling of processes in the context of services.

Typical process-oriented requirements are listed in Table 4.3. They stem from the way SMF is used by stakeholders and entities related to it.

The concept of roles amounts to a requirement of being able to refer to a set of tasks independently of the specific persons or other kinds of actors carrying them out.

Table 4.3 Process-related requirements

Definition of processes within individual stakeholders should be possible.
Definition of processes involving multiple stakeholders should be possible.
Processes involving both humans and automated actors should be supported.
Description of processes using roles should be possible.
SMF should support definition of access rights within and between organisations.
SMF should support definition of types of ownership to information.
SMF should support process traceability.

To cater for use within provider service environments, service modelling framework must support description of the workflow within an individual service provider. Such a description outlines temporal ordering of activities, including related needs discussed in Chapter 3. In addition, processes including multiple stakeholders need to be covered. Some subcontracting type examples of this were discussed in Chapter 3 in the context of TeleManagement Forum (TMF) service framework.

A future-proof framework must support defining processes for both automated and human actors. For the former, process choreography type definitions are viewed to become relevant for both intra-provider and inter-provider use.

4.5 Information Modelling–Related Requirements

Information modelling viewpoint collects requirements from the viewpoint of modelling disciplines and techniques. Requirements from information modelling viewpoint are listed in Table 4.4.

As we have seen, service models have multiple uses. On one hand, service model can be used for structuring interaction of a service management user with a service management system. On the other, it can also be used as a basis for automated communication between systems involved in service management and elements participating to service control. In general, the two types of information models do not need to be identical.

Further uses for service modelling include structuring of information for presentation in user interfaces, and in human-to-human interaction. Again, information modelling for human-oriented interaction can be different from one oriented towards automated processing. It is nevertheless useful if different kinds of models are based on the same concepts. Thus, the SMF should be a basis for service management, allowing its use by

Table 4.4 Information modelling–related requirements

It should be possible to use SMF for devising representations understandable to human users.
SMF should support mapping to formal representations.
SMF should be flexible with respect to ontologies.
SMF should support generic concepts as well as domain-specific modelling.
SMF should support principles of meta-modelling.
SMF should accommodate use of policies.
SMF should support SoA type modelling of services.
SMF should support modelling of state transitions in distributed services.

both human users and mapping to information structures relevant to automated communication.

Service management should not be restrictive regarding ontologies or entity taxonomy and relationship models. It should be possible to devise ontologies according to need, in addition to using single ontologies for multiple services and domains. This requirement seeks to accommodate both industry-wide modelling activities such as Shared Information/Data (SID) and more web services-oriented modelling.

It should be possible to present SMF as consisting of a common part, as well as more detailed sub-models devised for a specific purpose. The common part of the model ensures the overall consistency of the model, whereas domain-specific models provide the necessary 'machinery' for addressing particular needs of individual technology environments.

In the list, there are some specific requirements relating to specific paradigms currently in vogue. The paradigms in question are SoA and policy-based management (PBM). The optimal application of both of them to service management is presently being studied. Including them into the list is not meant to represent them as final truths, but rather as technologies that, according to current understanding, bring added value to service management.

4.6 Stakeholder Type–Specific Requirements and Concerns

Multiple stakeholder viewpoints are useful for studying requirements and concerns related to a system or an architecture from different perspectives. It enhances concreteness of the analysis without overly losing generality. An alternative approach is to dispose of stakeholders altogether and only discuss process areas. The latter approach is valid but considered to be too abstract for the present purpose, since the overall topic area is very broad. Subsequently, we shall stick to stakeholders in the present discussion. This should not be a serious limitation, since we shall not be discussing business models here. We shall describe stakeholders in a procedural sense, bringing them a step closer to process-oriented approach.

We shall next study some typical stakeholder-type–specific requirements and concerns. To stay on a generic level, we need to consider stakeholder types rather than individual stakeholders. To start with, we need to identify which types are relevant for the present purpose.

The following stakeholder classes are considered here.

- End-user/subscriber
- Service provider
- Connectivity provider
- Enabler provider.

We shall provide descriptions of each stakeholder class as well as examples in connection with stakeholder views as mentioned in following text.

In some stakeholder analyses such as MobiLife (2006), a larger set of stakeholder types have been identified. Here, we strive to keep the number of stakeholders as small as possible. One reason for this is that present analysis seeks to be as generic as possible, whereas, for example, MobiLife analyses an architecture for mobile applications.

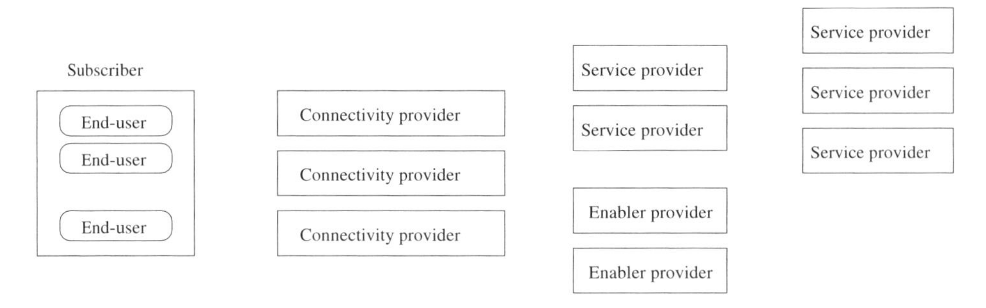

Figure 4.4 Illustration of stakeholder types

The list of stakeholder types above facilitates some of the most important aspects from the viewpoint of service modelling, including separation of subscriber and end-user, and separation of connectivity provider, service provider, and enabler provider from each other.

The 'big picture' of the relation of the stakeholders with respect to each other is illustrated in Figure 4.4. End-users can use multiple connectivity providers for accessing a particular service, and use services of enabler providers in the process. Service providers can aggregate services from other service providers. Where applicable, direct contractual relation exists between subscriber and provider. Please note that this figure is a special case of the earlier Figure 1.1. Specific examples of the stakeholders will be provided in Part Three.

4.6.1 End-user/Subscriber

Subscriber is the party that has agreements related products and constituent services. The contracts themselves may assume many forms, and could in principle be dynamic and based on automated mechanisms. End-user is the party that uses the services belonging to the product. End-user may be a subscriber, or subscriber could be the employer of the end-user, for example. In the latter case, we can speak of aggregate subscriber.

Typical end-user/subscriber stakeholder class requirements for SMF are listed in Table 4.5. It is useful to pay attention to the fact that there may be simultaneously relations both between the service provider and aggregate subscriber on one hand, and between service provider and individual subscribers on the other. In the former case, the

Table 4.5 End-user/subscriber stakeholder class concerns

Support for aggregate relation to other stakeholders by aggregate subscriber.
Support for one-to-one relation to other stakeholders by end-user.
Support for managing end-user-related information by the subscriber.
Support for expressing of subscriber-level preferences, including privacy-related ones.
Support for expressing end-user level preferences.
SMF must support means of making service quality predictable.
Architecture should not exclude future use of brokering-based mechanisms for service or access selection.
Service orchestration at communications endpoint should be possible.

relation is typically of aggregate nature, whereas in the latter case, it typically relates to supplementary subscriptions, management of personal preferences, or self-provisioning.

When subscriber and end-user are separate actors, it is usually necessary to have separate sets of requirements for both.

Service quality relates to both subscriber and end-user. In the former case, service quality refers to SLAs between the subscriber and a provider party. In the latter case, the predictability of service quality for individual service usage sessions is important for the usability of the service.

The reference to brokering mechanisms refers to the ability to select providers using automated negotiation means. Brokers have been considered related to access provider (Personal router whitepaper, 2006) and service provider selection (MobiLife, 2006), for example.

4.6.2 Service Provider

Service provider is here a class of roles related to providing of services – packaged as products – to other stakeholders. Providing of connectivity is considered separately in the next subsection.

Service provider may belong to one of the following classes:

- Service aggregator. This class of providers aggregates other providers' services, which in turn may be service aggregators or elementary service providers
- Elementary service provider
- Content provider is a specialised type of service provider
- Non-managed service providers offer services using lightweight platforms
- Service brokers. This is still partly a research topic, and the reader is referred to CADENUS, MobiLife, and Ambient networks projects (Ambient, 2006; Cortese *et al.*, 2003; MobiLife, 2006).

Generally speaking, connectivity provider can also be viewed to be a service provider. In view of the importance of service distribution, we treat it as a separate stakeholder class in this book.

Service provider may or may not have contractual relation to a subscriber prior to service use. In the latter case, service providers need to be discovered by the end-user. Discovery may be handled by other providers, and the process of discovery may be invisible to the subscriber and the end-users.

Services can be supported by a managed platform or provided in peer-to-peer fashion as illustrated in Figure 4.5. The latter case is used as an example of non-managed service providers. The degree to which service performance is predictable depends on whether they are provided on a managed or non-managed platform. Both types of services can coexist and be combined and used simultaneously with each other from the viewpoint of the end-user.

Description of the entities in Figure 4.5:

- ServiceProvider: stakeholder who provides services
- ManagedSP: provider of managed services, a subtype of ServiceProvider
- Peer-to-peer SP: provider of peer-to-peer, non-managed services, a subtype of Service-Provider.

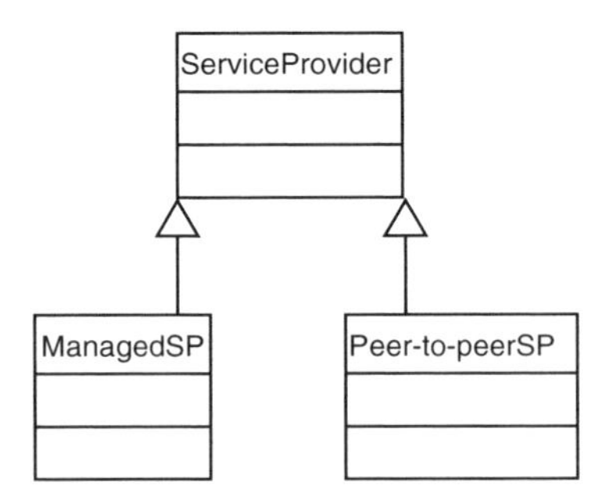

Figure 4.5 Managed and peer-to-peer stakeholders

In view of the envisioned increase in importance of dynamic service composition, framework must support cases in which the ultimate responsibility of the overall service performance lies within subscriber or end-user, and not with a service aggregator. For example, peer-to-peer services may fall in this class. Also in this case, it should be possible to define and supervise responsibilities of performances of constituent components and partition responsibility accordingly to aggregation-related roles.

Typical service provider stakeholder concerns are enumerated in Table 4.6.

Product-related data is relevant for providers of managed services and relates to business-related data. Possible compensation mechanisms for peer-to-peer services are not considered here. It should be possible to create customer-specific or customer class–specific variants of products and services. Service management framework should support grouping of end-users for the purpose of managing mass-market services.

SLA and Service Level Specification (SLS) are needed between providers of managed services, as well as between providers and subscribers. Conceptually, non-managed services can also be viewed as being associated with an SLA. We shall discuss this further in the next chapters. Service management framework must also support mechanisms and parameters relating to inter-provider operations.

Table 4.6 Service provider stakeholder class concerns

Product-related data needs to be supported both in service provider–end-user and service provider–service provider interfaces.
Support for SLAs and SLSs is needed. Liability limitation should be possible. Product, service, and service component life cycles should be supported. Self-provisioning of end-user services should be supported. It must be possible to provide predictable service quality to the end-user. Parameters relating to connectivity between service providers must be supported by SMF.
Grouping of end-users should be supported for mass distribution services. Tailoring of products and services to customers should be supported.
Service orchestration performed by service provider should be possible. Use of service registries should be possible.

The framework must support definition of end-user service performance targets by the service provider, including service quality related ones. The definition itself may be detailed or broad, depending on the need. For example, advanced multi-service networks such as Universal Mobile Telephony System (UMTS) support detailed service quality definition, whereas for WLAN service quality could be defined as a maximum, and peer-to-peer service be provided virtually without guarantees.

4.6.3 Connectivity Provider

Connectivity stakeholder type provides IP connectivity to end-users. Connectivity provider may or may not have contractual relations with subscriber. Where such relations exist, they may or may not be related to service provider relations. An example of the case where subscriber has a relation with connectivity provider is GSM or UMTS network, where subscription is associated with a SIM or Universal Integrated Circuit Card (UICC) smart card. Continuing the example, the service in question may be provided by the mobile operator group or by an external party. Peer-to-peer connectivity takes place between end-users without mediation from session control such as IMS. Peer-to-peer connectivity can mean either direct link layer 'ad hoc' connectivity or an 'overlay' multi-hop connectivity using other networking technologies. Overall, connectivity provider types include the following:

- Access network provider providing first-hop connectivity
- Backbone network provider offering connectivity between access networks and services
- Access broker supporting selection between multiple access providers. This is again a research topic, and the reader is referred to (Ambient, 2006; Cortese *et al.*, 2003)
- Peer-to-peer connectivity.

In the above list, the same stakeholder may operate both access and backbone network. In this book, we shall concentrate on 'last mile' connectivity providers. End-user may itself be an access broker, as described in (Personal router whitepaper, 2006).

The last type of connectivity may be based on Near-Field Communications (NFC) technology such as 802.11 in ad hoc mode between communications endpoints, or Ultra-Wide Band (UWB) communications, for example. Also, technologies for multi-hop ad hoc communications are being developed. In the case of peer-to-peer communications, endpoints do not always need a separate connectivity provider.

Both managed and peer-to-peer connectivity can be used simultaneously by an end-user, also as part of a service that is perceived as being homogeneous by the end-user. Generic types of connectivity are illustrated in Figure 4.6. The type of connectivity in question affects the types of services that can be supported. Also, as we shall see later, connectivity is in some cases intimately related to the composition of the service itself.

Description of the entities in Figure 4.6:

- Connectivity: connectivity between two stakeholders
- ManagedConnectivity: connectivity managed by a stakeholder. Associated with service level definition
- Non-managedConnectivity: connectivity that is provided without guarantees.

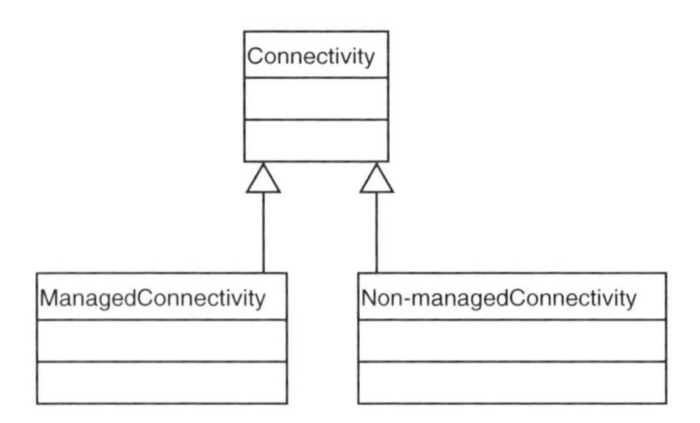

Figure 4.6 Connectivity types

Table 4.7 Connectivity provider stakeholder class concerns

Support for service quality allocation according to requirements and characteristics of services.
Support for multiple, heterogeneous simultaneous connections to single end-user.
Purely negotiated, purely provider-provisioned, and mixed service quality support allocation modes should be supported.
Provisioning based on user class, service provider, and service type should be supported.
Both static and dynamic service provisioning should be supported.
Both managed and non-managed bearer should be supported.

Some networks may pose limitations for direct use of 'overlay' type peer-to-peer networking. To be able to use peer-to-peer networking, the endpoints need to discover each others' Internet Protocol (IP) addresses. Different kinds of registries can be used for this purpose.

Typical connectivity provider stakeholder concerns are listed in Table 4.7.

The basic plot above is that it must be possible to allocate suitable connectivity for services where relevant. For example, in mobile networks it is important to use the licensed spectrum as efficiently as possible, whereby it is beneficial to specify connection parameters in greater detail than in unlicensed technologies such as WLAN.

From the viewpoint of supporting a broad range of technologies, different provisioning modes need to be supported. Some systems support service quality support instantiation on per-session basis, whereas for others, granularity of service quality allocation can be based on static information such as provider identity or protocol number.

Cellular connectivity and Asynchronous Digital Subscriber Line (ADSL) local loop are examples of managed bearers, whereas aforementioned 802.11 in ad hoc mode is an example of non-managed bearer.

4.6.4 Enabler Provider

Enabler provider is a relatively new stakeholder class as a separate entity, which provides commonly used functionalities, or enablers, for service providers as well as end-users. Enabler providers are expected to be an important ingredient for distributed services. Enablers can also be used as components in peer-to-peer services. The 3GPP Generic

Table 4.8 Enabler stakeholder class concerns

Discoverability.
Adequate connectivity.
Life cycle considerations relating to enablers.
Knowledge of endpoint capabilities.

Authentication Architecture (GAA) is an example of platforms using which enabler type services can be provided. The work done within Liberty alliance allows federation of authentication between service providers. A role called *Identity Provider* (IdP) is described in Liberty framework. On a more forward-looking note, the role of generic context providers has been studied, for example, in (MobiLife, 2006) and (Zuidweg *et al.*, 2003). Context providers can analyse context information and provide added value based on it. Terminal management is considered to be an enabler here as well.

Enabler stakeholder related concerns are listed in Table 4.8. The life cycle considerations referred to earlier relate to service enabler platforms such as OMA Service Provider Environment (OSPE) in the form of requirements for management of components. Typically, enabler provider environment needs to have high availability, and support hot swap of service components, for example.

The last requirement stems for classification of terminal management as an enabler.

4.6.5 *Stakeholder Interrelationships*

Let us next consider aspects of provider interrelationships by describing how different stakeholder types are related to each other. An overall 'genealogy' of stakeholder types is illustrated in Figure 4.7. The diagram shows two kinds of relations, namely, stakeholder types being special cases of other stakeholder types on one hand, and a stakeholder type aggregating other stakeholder types on the other. The aggregation relation here relates to services rather than business aspects. For example, the diagram indicates that

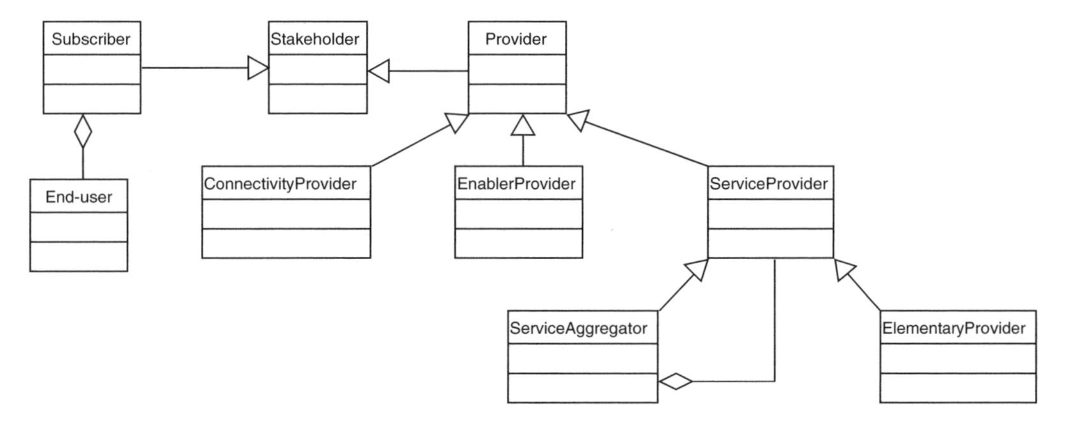

Figure 4.7 Stakeholder interrelations

a service aggregator can aggregate other aggregators, allowing for recursive construction of services. This is to be read in the sense that for a specific service, a service provider can aggregate other providers' services.

Description of the entities in Figure 4.7:

- Stakeholder: parent class for stakeholders
- Provider: parent class for provider-type stakeholders
- Subscriber: Stakeholder who has agreements with Providers
- End-user: uses agreements of Subscriber
- ConnectivityProvider: provides managed connectivity, subtype of Provider
- EnablerProvider: provides enablers for services, subtype of Provider
- ServiceProvider: provides services, subtype of Provider
- ServiceAggregator: constructs services by aggregating other services for Service-Providers
- ElementaryProvider: provides services that are not dependent on other services.

In the general – and most forward-looking – case, it is assumed that the set of providers can be dynamic, and that providers can be located on the basis of a registry mechanism. Also, the case of static inter-relations needs to be supported by the SMF.

The relations between individual stakeholders may be based on traditional contracts or on automated transactions. In both cases, technical content of the agreements must be sufficiently detailed. The SLS concept of CADENUS project is an example of automation of a concept previously related to traditional contracts. The framework needs to support translatory relations so that agreement with a stakeholder implies access to other providers' services, too.

Further examples of provider inter-relations will be provided in Section 5.2 as well as in the context of use cases provided in Part Three.

4.7 Summary

We started by describing the notation to be used in modelling and proceeded to analyse requirements for service modelling by using stakeholder classes. We concluded the discussion with a discussion about inter-relations of stakeholders.

The set of requirements identified here is a result of the viewpoints chosen for analysis. The intention is not to convey an image of complete requirement analysis, but rather a shortened representation of the background provided in Part One. Indeed, it would even not be possible to cater for all possible environments. Subsequently, the requirements description is best read as a partial best current practices description.

We shall have more to say about inter-relations of stakeholders in Chapter 5, and will present service framework that can be used for information exchange in Chapter 5. Modelling patterns described in Chapter 7 will provide a partial answer to the set of requirements, and the utility of the requirements for a particular environment needs to be assessed during design and implementation of the ultimate service model.

4.8 Highlights

Ten things to remember from this chapter:

- The requirements described are towards a full-service model in a production environment.
- Requirements are analysed using generic concerns and stakeholder-specific concerns.
- Stakeholder inter-relations must not be specific to any particular value net.
- Modelling must cater to both managed and non-managed services.
- Peer-to-peer services are used as an example of non-managed services.
- Service functionalities operated by enabler providers can be used in non-managed services.
- Connectivity can be managed or non-managed.
- Service quality allocation relates to service providers and connectivity providers.
- Service provider can aggregate other service providers' services.
- Peer-to-peer service providers need to be modelled, too.

5

Management Framework

In this chapter, we shall describe the overall framework within which service management takes place. It provides context for service management in general, and for using service modelling in particular. It expands and formalises the definition of service modelling and its context provided in the introduction. Service modelling needs to be formulated in such a way that management framework processes can make use of it in the best possible manner.

We shall describe an overview-type description of the management framework here, instead of explaining a complete process description such as the one described in Enhanced Telecom Operations Map (eTOM). However, we shall study processes closely related to service modelling in more detail.

We shall start with a description of the framework, and proceed to discuss assumptions of the inter-relations between stakeholders. We shall conclude this chapter with a summary of the relation of the described framework to other frameworks.

5.1 Description of the Framework

The management framework is described in three parts. The first one describes different views regarding modelling management tasks. The second part describes our management framework within which operability-related processes take place. The third part is specific to service management, and describes the phases relating to life cycles of services. We shall discuss contractual issues after that and conclude with a summary.

5.1.1 Views to Modelling Management Tasks

On the most general level, management framework needs to account for all the processes associated with management. Because of the complexity of issues involved, application of multiple views turns out to be useful once more.

Perhaps the most well-known set of views to management framework is the following one:

- Business management
- Service management

- Network management
- Element management.

Business management is responsible for products, service management for services, network management for network-wide management tasks, and element management for the management of individual network elements or classes of network elements. Each of the layers can be associated with strategy, development, and operations-type activities. Obviously, the classification given above is rather crude and cannot be expected to account for all the necessary details. It has served a purpose in facilitating the structuring of activities.

Perhaps the biggest challenge for the application of the above mentioned framework stems from the diversification of business environment. The number of stakeholders is increasing, and it is expected that there will be specialised actors for specific kinds of activities. Nevertheless, the framework is still largely applicable to processes within individual stakeholders. It is useful to note that many operator processes involve multiple layers.

There are certain things to take into account in using the framework in a modern environment. Firstly, the four layers should not be viewed as being completely separate from each other. The processes within the four layers are increasingly intertwined, whereby the framework only provides a rough map. Activities within the TeleManagement Forum (TMF) New Generation Operations Support Systems (NGOSS) programme provides good examples of this. Secondly, the use of the framework between enterprises is changing. Earlier, the most important forms of inter-enterprise communication took place in business management level. In the world that is heading towards web services and Service-oriented-Architectures (SoAs), deeper integration is called for.

Another set of views, more geared towards service quality management approaches management processes from the viewpoint of services. The following activity types can be identified:

- Capacity management
- Service quality support management
- Optimisation.

Capacity management is concerned with assessing and implementing network capacity in view of the needs of the subscriber base and service palette available. Service quality support management is concerned with allocating network capacity between services and end-users. Optimisation is concerned with ensuring that service quality support parameters best match traffic profiles at given times. From the viewpoint of timescales associated with the different tasks, capacity management clearly operates on longest timescales, and is associated with strategies. Service quality support management, in turn, could be viewed as tactics, trying to optimise actions in view of changing situation, given the limits set by strategy. Optimisation operates on a shorter timescales still, but is often a very important part of an efficient operation of mobile networks, for example, in today's competitive environment (Laiho and Acker, 2005).

On a high level, a degree of commonality can be perceived relating to products, services, and resources. In the most general case, each of the three areas can be viewed by being associated with strategy, tactics, and implementation. Indeed, this basic structure – albeit with different naming – is also reflected in the construction of the eTOM process map that is shown in Figure 5.1. As discussed in Part One, the three areas are inter-linked.

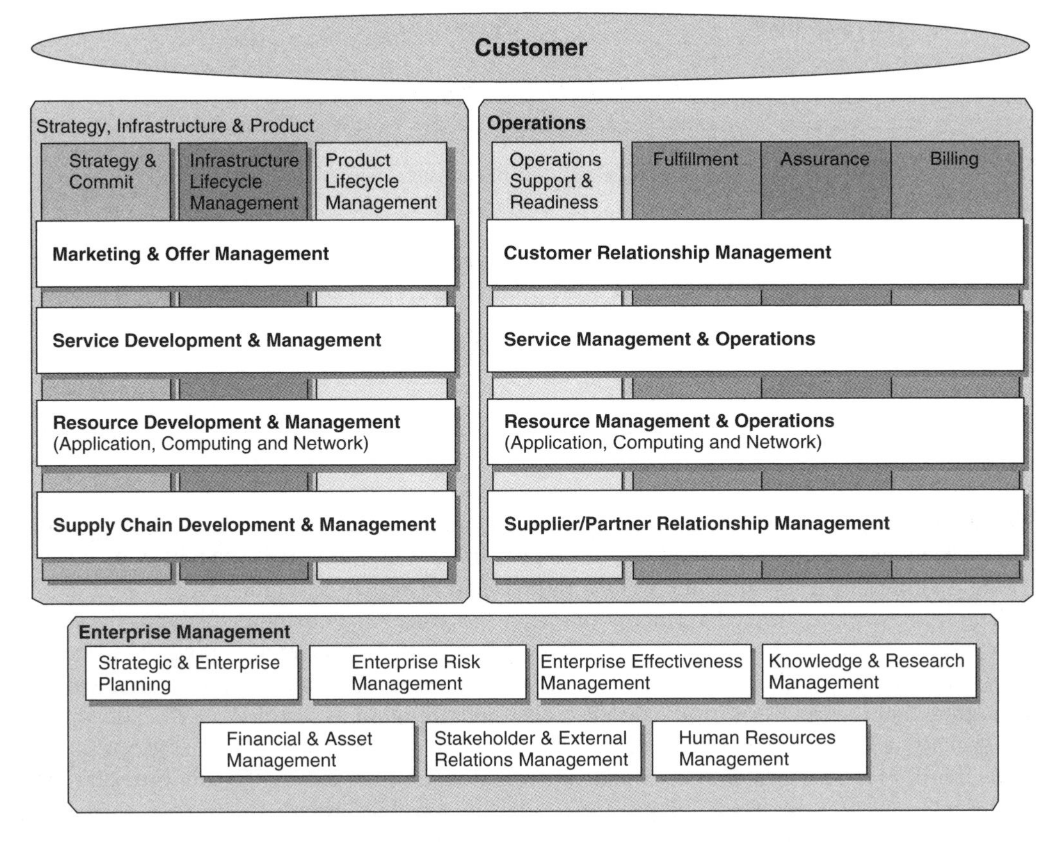

Figure 5.1 Layer 2 view to eTOM model. (Reproduced by permission of TeleManagement Forum)

In view of the emphasis on service management in the rest of the book, it is important to observe that service management must link to product and resource management processes and accommodate input from operations relating to these process areas. In the eTOM model illustrated above, product life cycle management cuts across service and resource process areas. In linking processes related to products, services, and resources, service modelling plays a central role.

Furthermore, it is important that the management framework supports multi-provider business operations and does not restrict the types of value nets to which a provider can participate. In view of the trend of 'deeper' business integration trend discussed previously, inter-enterprise operations should be supported on multiple levels.

Finally, a note about the impact of web services and distribution on management paradigms. We are still living the early days of SoA deployment and web services, and the final shape of things is yet to be seen. It has nevertheless been noted that web services environment lends itself naturally to a paradigm where web services management layer can be viewed to be positioned between the end-user and 'raw' web services (Hill, 2004). This approach means, in practice, application of policy framework to web services.

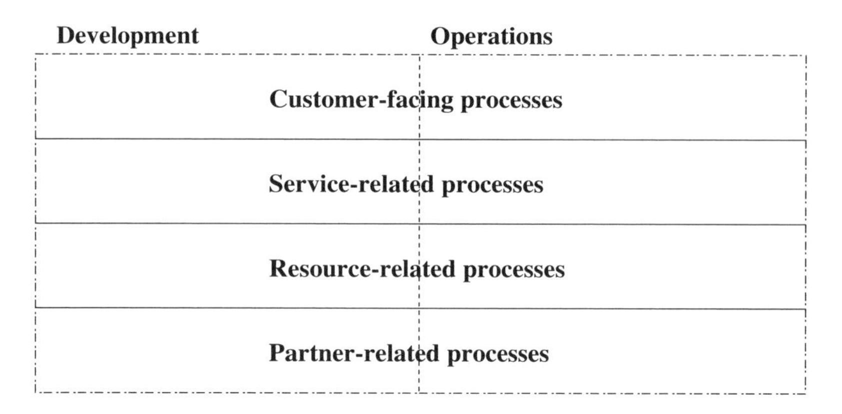

Figure 5.2 Management framework

5.1.2 Management Framework

We shall now proceed to describe a management framework that we can use for framing our subsequent discussion about service modelling. The framework can be viewed as being a simplified version of the eTOM model, and is shown in Figure 5.2.

The framework is divided vertically into four process areas: customer-facing processes, service-related processes, resource-related processes, and partner-related processes. In the horizontal direction, the framework is split into development and operations halves. In customer facing processes, for example, customer relation is formed in the 'development' half, and Customer Relationship Management (CRM) takes place in 'operations' half. Similarly, services are designed and created on the left half, and operated on the right one. Resource-related processes on the left half relate to making sure that infrastructure requirements are up-to-date with respect to other development half process areas. Finally, partner-related development processes relate to assessing partners and forming relations with them, whereas partner-related operations relate to partner management.

The management framework described can be used for describing the tasks carried out by provider-type stakeholders in our value net. Thus, services may be provided to end-user–related subscribers, or to other providers. An issue we shall return to, same entity may be perceived differently by different stakeholders. For example, capacity purchased from connection provider may be viewed as a resource by a service provider.

We shall illustrate the use of the framework with three examples: product creation, capacity management, and service optimisation. The positioning of examples with respect to management framework is shown in Figure 5.3. The positioning of the examples in the framework is dependent on assumptions to other processes, and we shall discuss this in the following text.

Product creation

We shall describe service life cycle related processes in more detail in Section 5.1.3, whereas we only provide a short summary here.

The feasibility analysis part of product creation involves analysis of business feasibility of a new product (customer-facing processes), analysis of feasibility in technological

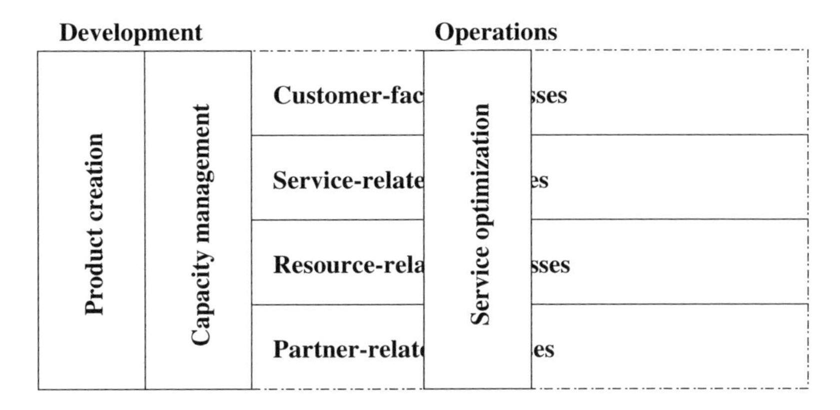

Figure 5.3 Positioning of examples in management framework

terms (customer-facing, service-related, and resource-related processes), and partnering feasibility analysis (partner-related processes). Overall, feasibility analysis makes use of portfolio of own and partnered products and services, and needs to consider resource development needs related to different options.

Actual product creation part involves definition of composition of product in terms of other products and services. Where relevant, new services are designed and implemented. As a part of design collection, which is collection of business and technical data relating to product and constituent services, is defined. These include performance counters for services, and usage statistics for services and products.

After product creation is complete, related functionality is parameterised, configured, and managed in the operations half of the framework.

Capacity management

Capacity management needs to make sure that sufficient resources exist for operating services according to performance targets and agreements towards external parties. Capacity expansion decisions may be triggered, for example, by increase in service usage, increase in the number of subscribers either locally or globally, or because of inadequate service quality.

It is easy to see a link to an example of product creation here: data collection defined as part of product development is used as an input to capacity management.

Optimisation

Optimisation is a set of tasks related to making the best use of available resources for a given set of services. Thus, it belongs to the operations half of the management framework. For example, the service quality support parameters relating to a particular type of bearer may be modified to enhance service quality in the face of increasing usage of resources in particular location.

In order to function, optimisation needs to have the right counters in place. Again, link to product creation example is of help here.

5.1.3 Service Life Cycle

Next, we shall concentrate on framework for service life cycle management, while making notes about linkage to more general context, too. Parts of service life cycle fall within the product creation example.

Service life cycle can be viewed as constituting a further viewpoint to the stakeholder tasks, being related to tasks carried out during the entire lifespan of products and services.

For our discussion, we shall assume that there is a service provider who is responsible for a service, which may be used as a part of a product. The service may make use of other services, and may be used, in turn, as a part of another service. It is further assumed that resources and/or constituent services can – but need not – be represented as components, as is the case in Open Mobile Alliance (OMA) Service Provider Environment (OSPE). Services and components can be associated with sets of parameters.

From the viewpoint of service provider, service management process may be viewed as consisting of the following parts:

- Requirement specification
- Service creation
- Service configuration and provisioning
- Service operation
  - Service performance assessment
  - Service optimisation
- Service retirement.

Requirement specification is based on business assessment, which we briefly described in connection to the product creation example. As described in (Service Framework, 2004), requirement specification can involve business and technical roles, and produces a technical specification of the service in question. In the process, solution-type specification work may be performed, taking into account existing services and agreements towards other providers. This phase includes the definition of target service performance levels.

Service creation can include implementation of new service components, parameterisation of reusable components, and taking care of necessary actions relating to possible third-party service components. Making necessary configurations for connectivity towards users of the service as well as between the components of the service is also part of service creation. Definition of the collection of necessary metrics for the operations and optimisation phases needs to be configured at this stage, too. Typically this involves Key Performance Indicators (KPIs) and Key Quality Indicators (KQIs) relating to service performance. Service modelling is assumed to be a part of service creation phase, producing information about interconnections among products, services, and resources.

Up to this point, the tasks have been carried out in the 'development' half of the framework.

Operations part makes sure that the service is adequately linked to resources and that the connectivity parameters are set correctly.

After necessary technical configurations have been done, the service can be provisioned to end-users. In addition to enabling the service for end-users, this stage may involve making end-users aware of the existence of the service. For web services using a registry, the service may need to be registered to be discoverable.

Service operation is responsible for managing the service according to planned performance levels, and ensure that possible corrective actions are carried out. Analysis of service performance compared to planned level plays a central role here. Service performance estimation makes use of service models, enabling service impact analysis.

As we saw in the examples, service usage metrics are used for resource development activities, too. In addition to examples about optimisation provided, changing the parameterisation or composition of the service may be performed in such a way that it better fits to changes in business environment or resources available for operating it. Thus, optimisation may relate to service composition, too, in which case it needs to be brought back to 'development' half of the diagram.

Service retirement includes activities relating to taking a particular service out of use. It must be made sure that there are no longer active users on the service, and it needs to be removed from relevant registries. Components that are no longer used by services can be disabled.

The phases stated above relate to the life cycle of a single service. In addition to supporting this, modern service management systems also cater for service portfolio management. In such systems, reusable components can be effectively utilised. Also, representation of the linkages and dependencies between individual services and components is important. Support for solution-oriented service definition means the ability to review capabilities of existing services and to reuse existing services as building blocks in constructing new ones. The system should allow operation of multiple versions of the same service. Service management is also assumed to be able to facilitate efficient communication between business requirements, services, and resources. In addition to supporting dependencies between different levels, service management system is also assumed to provide necessary mechanisms for business and resource development purposes.

Above, we have concentrated on life cycles of individual services. Generally speaking, other entities can also be viewed as having life cycles associated with them, including:

- Products
- Resources
- Components
- Configuration data.

As seen above, components are building blocks for services, but are not, necessarily, always services themselves. We shall not discuss other kinds of life cycles at length here. Some related information can be found in OMA OSPE work as well as in MobiLife project.

5.1.4 Service and Product Concepts

The basic concepts relating to services and products are shown in Figure 5.4. A product makes use of one or more services, which in turn make use of resources. Each of the levels is associated with a provider. The figure shows only the high-level inter-relation of the concepts, and more detailed modelling will be described later in this book.

Description of the entities in Figure 5.4:

- Product: an entity that can be sold to Subscriber
- ProductProvider: stakeholder responsible for packaging of Products

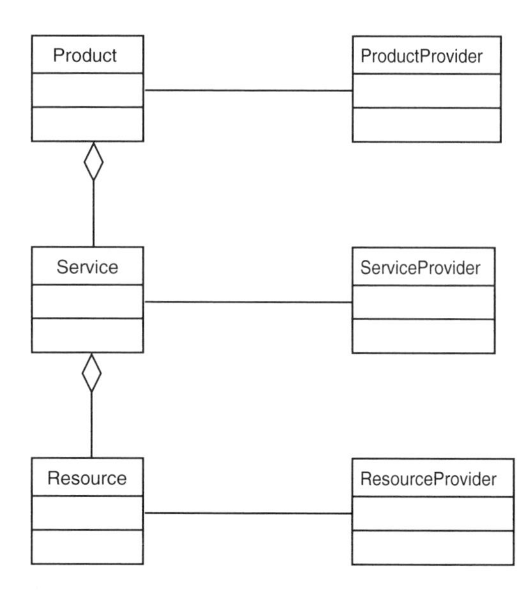

Figure 5.4 Basic concepts

- Service: an implementation of Product, associated with definition of acceptable behaviour
- ServiceProvider: defined previously. In this context, the stakeholder responsible for defining and maintaining behaviour of Service
- Resource: entities needed by Services
- ResourceProvider: stakeholder responsible for maintaining resources.

Earlier, we have defined a role as amounting to a set of tasks, which does not need to have a one-to-one mapping to persons. Multiple persons may be involved in a single role, and a single person may be involved in multiple roles. Same kinds of roles may coexist in different stakeholders, just as management framework can be applied to different stakeholders.

Different service management roles can be identified within provider-type stakeholders. These are illustrated in Figure 5.4. Some of the roles relevant to creating and operating services are shown in Figure 5.5. The reader is invited to take a look at (Service Framework, 2004) and (Koivukoski and Räisänen, 2005) for more discussion about roles. For our present purpose, it is sufficient to make a distinction between business-oriented roles and operations and management roles. We shall discuss specific roles in the context of service model in Chapter 7.

Description of the entities in Figure 5.5:

- Role: parent class for service management roles
- BusinessRole: Role responsible for business-oriented aspects of Products or Services
- OMRole: Role responsible for technical aspects of Products or Services, including maintenance
- Manager: supervisory Role
- SolutionDesigner: Role responsible for defining solutions using portfolio of Products, Services, and Resources
- Architect: Role designing technical implementations of Services

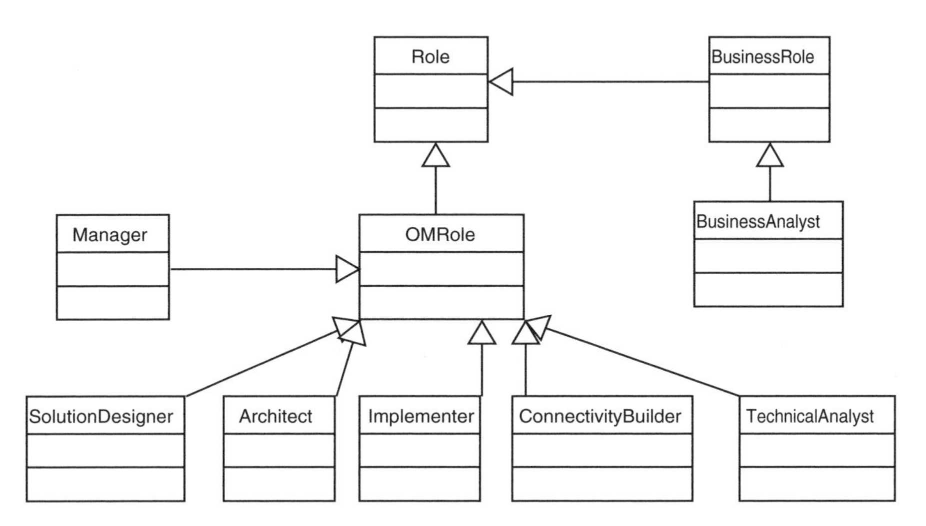

Figure 5.5 Examples of service management roles

- Implementer: Role implementing Services
- ConnectivityBuilder: Role configuring connectivity between stakeholders
- TechnicalAnalyst: Role performing technical analyses of service and role behaviour.

In addition to roles related to descriptions of tasks carried out by humans, roles can also be used for exposing aspects of services or resources from a particular viewpoint.

5.2 Assumptions about Inter-provider Relations

Next, let us discuss briefly the assumptions about inter-provider relations from the viewpoint of management framework. The management framework needs to take into account the way in which inter-provider relations as well as ones oriented towards customers are handled. We shall not describe procedural conclusions here, but merely describe aspects relating to management framework.

Contractual relations are associated with products that a stakeholder procures from the other stakeholder. We shall not consider the internal relations of stakeholder here.

It is assumed that a subscriber will have an agreement with one or more of the following parties: service provider, connectivity provider, and enabler provider. The situation is illustrated in Figure 5.6. It is assumed that at a minimum, it is enough that a subscriber has an agreement with one of the providers, and may obtain access to other providers' resources based on inter-provider agreements. For example, the subscriber may have an agreement with connectivity provider and get automatic access to enabler provider(s) and service provider(s). Explicit and implicit agreements between subscriber and providers have aspects which are visible to end-users, including impact on experienced service quality. Please note that inter-provider relationships are not shown in Figure 5.6.

Description of the entities in Figure 5.6:

- Subscriber: described previously
- End-user: described previously. In this context, the end-user can have direct relations to different Providers

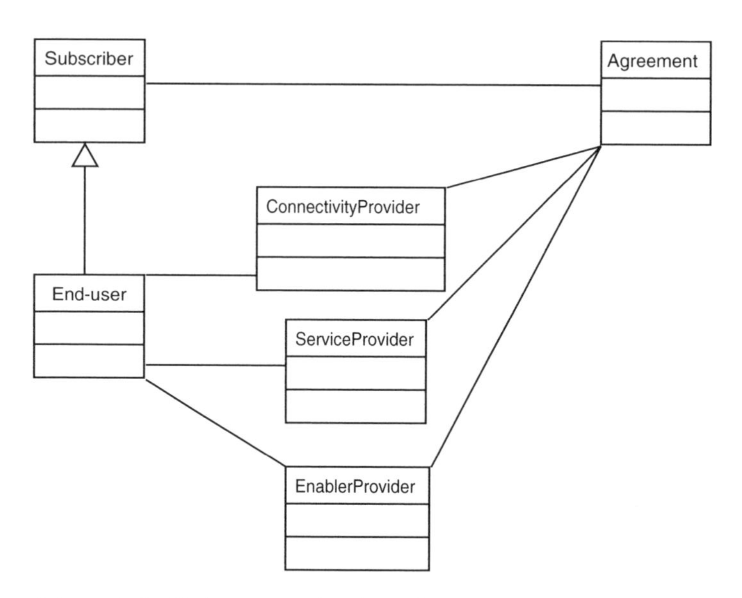

Figure 5.6 End-user centric view of agreement-related relations

- Agreement: formalisation of subscription relation between Subscriber and Provider
- ConnectivityProvider, ServiceProvider, EnablerProvider: described previously.

We shall make more inroads to modelling agreements in specific situations in Chapter 7 after we develop necessary conceptual basis for it. In particular, we shall make modelling of end-user–related associations more detailed.

The stakeholder framework described will be used as a 'high-level road-map' in assessing inter-provider relations. Role-based approach complements it by focusing on areas of activity that can be shared by stakeholders.

5.3 Relation to Existing Frameworks

The elements of service framework described above are in line with the principles of process modelling work done within TMF, but do not attempt to achieve similar level of detail. The basic motivation for describing the framework here is to describe which aspects of the state-of-the art survey in Part One are considered important for Part Two. We shall have more to say about relation to existing frameworks after presenting service model in Section 7.4.

The basic organisation of management framework into customer-oriented, service-, resource-, and supplier-related roles is the same as in eTOM. The relation of product creation to management framework is also analogous to eTOM. We have left out many important parts of eTOM in the interest of simplicity; for example, we have not considered billing process area at all here. In eTOM, assurance is a vertical process area in the operations half. We shall not use this construct in our analysis. Service topology defines the inter-relations of services and resources, and data collection is defined as part of product and service creation. On the other hand, we have illustrated optimisation-related processes in the operations part of the framework, which does not appear in eTOM.

The concept of service management roles is used basically in the same way as in (Service Framework, 2004). In the context of service management, roles represent areas of activity, which can be linked to information and process models. Roles will also be used for describing aspects of entities from a particular viewpoint.

Basic approach to information modelling is similar to that in Shared Information/Data (SID): information ownership can be related to processes, and fundamental relations relating to service management can be analysed using Unified Modelling Language (UML) models. There are some differences in details to SID within the model presented in this book, an issue we shall return to later on. There is a more fundamental difference as to the purpose of the model. SID is devised to be a reference model for communications industry, and seeks to be as complete and as consistent as possible. The model that we shall be presenting in Chapter 7 seeks to be internally consistent, too, but in contrast to SID, it is not meant to be a complete description of the problem area. Instead, we are describing building blocks for service models, illustrating specific areas of interest.

There are multiple reasons for opting for a more limited view than SID. Firstly, the SID model has been developed as part of limited-membership industry forum, and as such is not freely available at the moment. Secondly, the scope of the current book is on one hand wider than SID, and there is need to focus on certain issues more fully than what is done in SID. Regarding other existing models, either same kind of problems exist as for SID, or models are too simple to serve as a basis. Finally, service model needs to be ultimately tailored to stakeholder's technical environment.

As a consequence of these factors, the author has decided to present a dedicated set of views instead of a model that could be directly mapped to an existing model. References to some basic structures of SID will be made within Chapter 7, but there will not be a complete description of the relation to SID owing to reasons described in the preceding text.

5.4 Summary

We have described a management framework within which service management tasks are carried out. The management framework serves to provide a big picture of the operations within a stakeholder. In particular, service life cycle phases were described. We reviewed basic assumptions about inter-provider relations and concluded by describing relations to existing models.

Service modelling is used in management framework in different places. Service topology and associated process design are created in the 'development' half of the framework and used in the 'operations' half. For operations, service topology not only tells what kind of resources need to be configured for particular services, but also structures the configurations associated with services.

In the next two Chapters, we shall describe service framework used for technically describing services, and proceed to present service model in the form of patterns. The service framework is an entity that the management framework described above can utilise, and the same is true of a service model. Service modelling patterns provide building blocks for the latter.

5.5 Highlights

Ten things to remember from this Chapter:

- Service framework sets the stage for utilisation of service modelling.
- The division of operations into business management, service management, network management, and element management is often used.
- Utilisation of SoA ideas requires deeper integration between enterprises than what has been considered traditionally.
- Service management relates to other processes, such as capacity management, service quality support management.
- Life cycle stages can be identified relating to individual services.
- Collection of service quality metrics needs to be part of service creation.
- Life cycle thinking can also be applied to products, resources, components, and configurations.
- Stakeholders can be associated with different types of entities.
- Service management roles can be associated with different types of entities.
- Subscriber stakeholder used agreements towards one or more provider-type stakeholders.

6

Service Framework

We shall now describe a service framework which can be used for exchanging information relating to services. The information in scope is mostly technical, but also includes some business aspects. Relating back to management framework, product, service, customer, and partner oriented processes can use this framework for this purpose. Service quality and security are related to the framework, since they are often considered topics of importance for services. The framework that is described later can be used for other purposes, too.

We shall start by reviewing technical requirements and characteristics of services. The discussion in this chapter will help us to formulate service modelling patterns in a more concise manner by laying down some of the needed framework. The service framework can be utilised by service life cycle processes, and it is also of value for inter-stakeholder information exchange.

Some examples of the application of the framework are included in this chapter, and more will follow in the next one. More comprehensive examples will be presented in Part Three.

We shall start with an introduction and proceed to a discussion of service quality, presenting a framework for describing different aspects of services. Next, we shall discuss the security related issues framework and relate it to service quality framework. Analogously, security is discussed and related to the service framework. After that, we shall discuss requirements and characteristics of classes of services and conclude with a summary. The framework description is built on top of the service quality basis, because service quality support management is typically viewed to be challenging from the viewpoint of services.

The service framework presented in this chapter is based on (Räisänen, 2004, 2005) and is used for structuring information relating to both the service quality as well as security. Prior discussion about the relation of the service quality framework to security and other aspects such as mobility can be found in (Koivukoski and Räisänen, 2005). Further discussion about end-user services can be found in (Laiho and Acker, 2005).

The overall setting for the use of the service framework is that of multi-party service management: target service quality and security levels are determined by a provider party in service usage scenario, and service quality support is instantiated based on an explicitly or implicitly requested service quality level. The service quality range available to a particular user and particular session is defined as part of service quality management.

Thus, the service quality experienced by an end-user is affected by both service quality management of the connectivity provider and the service management of the service provider.

The connectivity provider and the service provider typically limit the range of service quality available to the end-user for resource reasons. Subsequently, instantiated service quality is not necessarily identical to the requested service quality level, but may be lower. The granularity of service quality differentiation varies from system to system. The security level targets can be basically set in the same way as for service quality, but there are fewer factors affecting the end-to-end security level than is the case for service quality. There is usually no need to limit the security level because of the resource availability reasons, but in some cases, legal aspects may affect the strength of encryption available.

The end-user is not typically interested in actual service quality parameters, and service quality negotiation may be handled by an application running in the communications end-point or middleware on behalf of the user, for example. On the other hand, the end-user may be interested in the service usage experience, as we shall see. This, in turn, has a bearing on the service quality parameters. The same kind of logic can be applied to security.

We shall not consider the means of supporting service quality or security here, but will describe a framework which can be used for managing service quality. For service quality, an interested reader can study, as an example, (McDysan, 2000) and (Armitage, 2000) for a summary description of Internet service quality support technologies. A summary description of mobile networks service quality can be found in (Laiho and Acker, 2005) and (Halonen *et al.*, 2003), for example. The security-related aspects and further references can be found in (Schneier, 1996), among other sources.

6.1 Introduction

Internet Protocol (IP) does not provide any service quality guarantees in the same way that circuit-switched systems do. The basic reason for this is that IP is packet-oriented rather than circuit-oriented, and each packet can be routed and forwarded independently of other packets belonging to the same service instance. The technologies which led to the development of the Internet were conceived with resilience considerations in mind, and led to design principles which were in part orthogonal to the service quality oriented design requirements. Allegedly, this fact led one of the industry luminaries to note tens of years ago that packet-based networks are not suitable for the delivery of telephony. Based on the world around us at the beginning of the third millennium, this judgement was a tad too hasty. There is some truth in the statement, though, in the sense that the addressing of service quality on the Internet has needed to focus on related issues and has benefited from the development of supplementary technologies.

As an example of this, higher-layer protocols or application-level mechanisms are needed to implement reliability with IP. On the other hand, mechanisms on both the lower and the higher layers other than IP can be employed for providing and controlling service quality. The challenge lies within the fact that requirements regarding service quality vary from one class of service to another, and some services are more amenable than others to service quality defined freely by a provider. We shall present examples of both these extremes later on in this chapter.

Analogously to service quality, security was not one of the design goals of the Internet. In circuit-switched networks, content was transferred in a dedicated physical network, providing a natural, albeit simple, security mechanism. As we have previously seen, the trend is towards multi-service networks capable of transferring any digital content. As more and more services are operated on top of IP, security has become increasingly important. The reasons for this are many: monitoring of IP-based traffic is relatively easy, the monitoring tools are readily and freely available, and public awareness of the importance of security is on a higher level than it has been in the past. As with service quality, complementary services have been developed to support security on the Internet.

A framework for expressing technical requirements and characteristics of services is important for service modelling and management. Such a framework needs to be applicable to different networking technologies. We shall describe a framework for this purpose in this chapter. The service security will also be discussed in relation to the same framework. Service-related characteristics, will also be presented in the context of this framework.

6.2 Service Quality Framework

In view of the goal of developing a general, technology-independent framework, we shall step back and review service quality from a broad perspective.

We shall start by reviewing prior work, and proceed to listing requirements for the framework, based on the preceding introduction and literature survey. After that, we shall describe the actual framework and its use in service management processes. Examples of the use of the framework are provided in the next section. More general discussion about the classification of end-user services can be found in (Laiho and Acker, 2005) and (Halonen *et al.*, 2003).

6.2.1 Prior Work

In circuit-switched networks, service quality was easier to define than in packet-based networks, and basically amounted to a set of engineering parameters. For the Internet world, the definitions needed revision.

The International Telecommunication Union (ITU) defines Quality of Service (QoS) as follows (ITU-T Recommendation G.1000, 2001): *Quality of Service is the collective effect of service performances which determine the degree of satisfaction of a user of service.* The ITU recommendation proceeds to list the following four viewpoints to QoS:

1. QoS requirements of customer.
2. QoS planned by provider.
3. QoS delivered by provider.
4. QoS perceived by an end-user.

Above, the customer and provider viewpoints are separated from each other, as are planned and delivered service quality. ITU has produced a classification of end-user services into four categories in terms of end-to-end latency (interactive, responsive, timely, and non-critical) and two categories with respect to sensitivity to bit errors (error-tolerant and error-intolerant) (ITU-T Recommendation G.1010, 2001). This concept has been useful in developing the understanding of service quality classifications.

The Third Generation Partnership Project (3GPP) has defined a QoS framework and corresponding architecture (3GPP TS 23.107, 2004; 3GPP TS 23.207, 2004) as an evolution of the General Packet Radio Service (GPRS) QoS framework. The QoS attributes belong to the properties of the bearer negotiated between the terminal and the network, and service quality is controlled by means of creating and managing bearers, and mapping of traffic onto them. The most important QoS attribute is traffic class, being one of conversational, streaming, interactive, or background class. The description of the other parameters such as end-to-end delay and bit rate parameters can be found, as in (Koodli and Puuskari, 2001; Räisänen, 2004), or (Laiho and Acker, 2005). The Release 5 (R5) of 3GPP architecture supports the dynamic linking of bearer characteristics to Session Description Protocol (SDP) parameters of IP Multimedia Subsystem (IMS) sessions based on Session Initiation Protocol (SIP), in addition to static provisioning of maximum service quality in earlier releases. Release 6 of 3GPP generalises support for session-based services. More information about 3GPP QoS framework can be found in Appendix A.

Internet Engineering Task Force (IETF) has described two service quality frameworks: Integrated Services (IntServ) (Braden *et al.*, 1994) and Differentiated Services (DiffServ) (Black *et al.*, 1998). IntServ is based on an explicit request for end-to-end service quality support from the network by the communication endpoint. DiffServ, on the other hand, is – in its original form – based on pre-configured service quality allocation at the network domain edge router. Adding the dynamic service quality allocation in DiffServ networks is possible, but it changes the basic philosophy of DiffServ a bit (Räisänen, 2003a). IETF also has other service quality related activities, addressing, for example, development of service quality signalling frameworks. We shall not consider them here since they are not pertinent to service quality framework. Appendix B contains some more details about DiffServ.

In addition to the 3GPP QoS framework and the two IETF frameworks, there are other, less well-known frameworks which we shall not discuss here. An interested reader is referred to (Räisänen, 2003a) or (Räisänen, 2004) for a longer survey.

Generally speaking, a QoS framework includes means of identifying an incoming or outbound traffic entity and allocating service quality for it. Service quality may be allocated based on negotiated service quality support (IntServ and 3GPP), or according to network policy (DiffServ and 3GPP). Using the terminology of (Räisänen, 2003a), this can be described as service quality support instantiation for a service instance. The criterion for service quality allocation policy can be the host or network of origin, service, or the user of the service. The modelling of service quality should not be limited to a particular service quality support framework such as IntServ or DiffServ, but should, rather, be sufficiently generic to lend itself to use in different environments.

One of the basic challenges in service quality in IP-based networks stems from the multitude of services operated on top of it. As we shall see in Section 6.5, service quality requirements vary from service to service. The easiest way to cater to all would be to use 'over-provisioning', that is, build so much capacity that it is sufficient for almost all imaginable usage scenarios. Unfortunately, this line of action is in some case prohibitively expensive, and as such, does not exactly support the goals of lean operations. Service quality support in real-world environments, therefore, boils down to using available resources intelligently, taking into account requirements of service types.

In order to request or allocate a particular service quality level, it is useful to differentiate between inherent and designed service quality requirements (Räisänen, 2003a). Certain services have inherent service quality requirements, the lack of fulfilling of which renders the service useless. The maximum end-to-end latency and packet loss limits for the voice signal of telephony is probably the most familiar example of these. For other kinds of services, certain aspects of service quality can be designed. The limiting of the maximum throughput of a browsing user is an example of this. The latter paradigm benefits from the ability estimate end-user perception of service quality. This discipline has the longest history in the field of the estimation of voice quality (TIPHON end-to-end Quality of Service, 2000; ITU-T Recommendation G.109, 1999), but the basic idea can be generalised and applied to other services as well (Bouch *et al.*, 2000; Räisänen, 2003a). Many services have limits for service quality parameters, which, when not met with, lead to notable deterioration in usefulness.

A good understanding of the characteristics of services is important for providing adequate service quality support, while utilising network resources in an optimal manner. The basic traffic characterisation methods include measurements and traffic modelling. Examples of applying these methods to wireless environment abound; a couple of examples can be found in (Heckmann *et al.*, 2002) and (Klemm *et al.*, 2001; Leung *et al.*, 1994), respectively. Measured or modelled characteristics can be used to determine traffic descriptors for services, and are also used as a tool in designing new systems such as multi-service mobile networks. They are also important for simulating and evaluating system performance (cdma2000 Evaluation Methodology, 2004). For our purposes, we assume that characteristics can be estimated in some manner.

To obtain a better understanding of what is needed of a service quality framework, it is useful to use a framework for describing services themselves, capturing the essential phenomenology relevant to service management. The framework should lend itself to analysing potentially heterogeneous properties of different aspects of a particular end-user service, yet be able to treat the end-user service as a single entity where necessary. The concept of IP multimedia session in 3GPP R5 is an example of such aggregation of concepts (Poikselkä *et al.*, 2004).

Let us discuss the phenomena considered essential prior to describing the framework. From the provider viewpoint, an end-user service has a set of parameters, including service quality oriented ones. There can be different variants of a service, related to different user groups and different access technologies, for example. Taking GPRS and wireless area local network (WLAN) as an example of the latter case, the maximum throughput can be different in the two variants. Thus, also characteristics can vary from variant to variant. The variant can potentially be instantiated with different sets of parameters.

An instance of service variant can consist of a number of flows, which may or may not be related to sessions, depending on the service in question. Different flows may be related to different applications, and have different characteristics and requirements associated with them. The service quality support can be instantiated for flows on a one-to-one basis, or based on aggregation criteria. In the latter case, flows of different types are treated as the basis for service quality allocation. The DiffServ framework makes use of this approach, for example.

A four-level framework was proposed in (Räisänen, 2003a), consisting of aggregate service, service variant, and service event levels (Figure 6.1). Service events can be further classified into service event types for service quality allocation purposes. The goal of

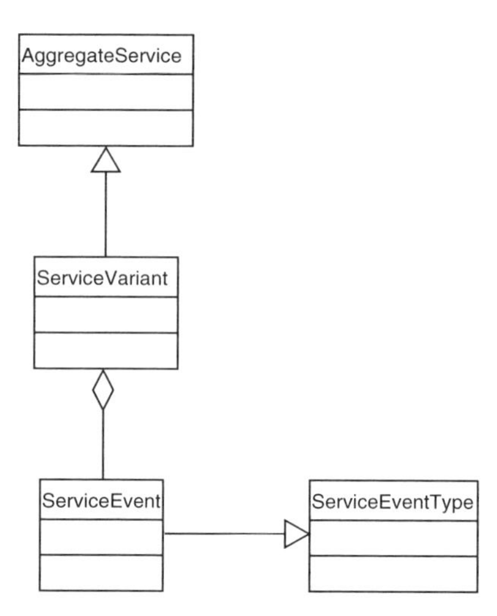

Figure 6.1 An illustration of service modelling concepts from (Räisänen, 2003a) using Unified Modelling Language (UML)

having such a framework is the ability to model multiple variants of an end-user service – for example, relating to different users – as well as the fact that different service variants may be composed of different sets of service events. Furthermore, service event types can be used for representing aggregated treatment where event-specific service quality support is not needed. Such a framework is also useful for describing advanced services, as we shall see below.

Description of the entities in Figure 6.1:

- AggregateService: most general description of a service, including different variants
- ServiceVariant: a variant of AggregateService, for specific end-user class or access technology, for example
- ServiceEvent: a related unit of traffic of homogeneous type relating to a specific aspect of a service, for example a flow or a transaction
- ServiceEventType: aggregation of ServiceEvents for service quality allocation purposes, for example.

Please note that in Figure 6.1, the 'diamond' symbol in UML denotes aggregation, so that, for example, a variant may be related to multiple events.

Using a framework such as the one described above, information about requirements and characteristics of end-user services can be recorded and exchanged between stakeholders. In the following section, we seek to use the framework shown in Figure 6.1 as a basis for a more formal and more general framework. Next, we shall list some requirements for the service framework. After that, we shall describe the entities of the framework in more detail.

6.2.2 Requirements

We shall use the requirements listed in Table 6.1 as a basis for adding more details to the service quality framework.

Table 6.1 Requirements for service quality framework

Modelling must not be tied to a particular service quality paradigm.
The framework must lend itself to mapping to technology-specific service quality support parameters.
The framework must allow for expressing service performance targets on per service, service provider, and per user basis. Also aggregation of users needs to be supported.
The framework must be suitable for different tasks relating to service management, including creation of a new service, changing its composition, as well as measuring its performance.
Also other processes such as resource development related ones need to be supported.
The framework needs to be able to describe multiple variants of a service.
The framework must be suitable for use in inter-provider negotiations.
The framework must support heterogeneous service quality requirements of entities belonging to a particular variant of an end-user service.
The framework must support the provision of the same service via multiple access technologies without having to know specific parameters of the access technology in use.
The framework shall support awareness of access technology.
The framework shall support distributed services.

Most of the requirements are based on the preceding discussion, but a few comments are in order. The second and third last requirements may seem to be in conflict with each other, but listing them both is meant to allow for both possibilities. Simple services may not need to know anything about the access technology, whereas advanced services may benefit from having at least some information of it.

6.2.3 Description of the Framework

The construction depicted in Figure 6.1 forms the basis for the framework. We shall describe in more detail the different entities and their relation to service quality the way that it is used in this book. After them, we shall describe the functions that span the different entity types.

From the viewpoint of service usage, parameters in the framework are used for negotiating service quality and instantiating service quality support for services, reflecting the result of the negotiation. Provider-type stakeholders typically use parameters for ensuring that service quality support matches the needs of the service, and not too many service instances are associated with a resource. Different service variants can be associated with different service quality support ranges. In addition, provider-type stakeholders can make use of the framework for exchanging information about planned and actual service quality levels, as we shall see in Chapter 7. Providers can also use the framework for business-to-business service purposes, such as leasing transport capacity. As we have previously seen, the purchasing of a service needs to be accompanied by a product definition.

The parameters have been classified into business parameters, technical parameters and service quality related parameters, the last category being divided into requirements and characteristics. The business parameters have relevance to product description, and technical parameters are also of relevance to intra-operator processes, as discussed earlier. The service quality parameters can, of course, be regarded as parts of technical parameters, and have been set aside for convenience and clarity. The set of parameters given below are not complete, but they should be considered as exemplary ones.

The concept of policy has been used for the expression of defaults on different conceptual levels. We shall return to this concept in this section after describing the entities relevant to service quality management. We shall start by describing relevant aspects of aggregate services, service variants, service events, and service event types.

Aggregate service

Aggregate service describes the aspects of a service that is common to different variants and instances of the service. Aggregate service levels can also be used for inter-provider information exchange related to end-user services, and needs to be supplemented with product description for procurement purposes. In what follows, typical parameters relevant to aggregate service are listed. All of them are not necessary for every service.

Parameters include:

- Business oriented parameters
  - Geographical coverage of the service
  - Validity period of the service
  - Service Level Agreement (SLA) parameters
  - Traffic Conditioning Agreement (TCA) parameters
- Technical parameters
  - Types of service variants
  - Means of selecting variant
  - Service Level Specification (SLS) parameters
  - Traffic Conditioning Specification (TCS)
  - Service-specific policy
- Service quality requirements
  - Availability
  - Retainability
- Service quality characteristics
  - Overall usage of the service.

Above, the 'means of selecting variant' refers to the logic by which adequate service variant is chosen when service is instantiated. Such logic can be automatically part of the operation of the network, or require invocation of special functionality for this purpose. The use of subscriber profile in 3GPP bearer activation in tandem with Access Point Name (APN) provisioning is an example of variant selection mechanism. The service quality support parameters associated with bearer instances may be different for end-users belonging to different QoS profile classes. For service quality characteristics, the usage parameters can be presented in different kinds of contexts, including geographical, temporal, and demographical.

There are many more details to aggregate services which are not central to the topic of this book and hence are not discussed here. For example, SLAs contain information relevant to reporting and means of measuring SLA fulfilment (SLA Management Handbook, 2001). These kinds of definitions are usually part of SLA definition and hence form a part of the aggregate service level.

Service variant

The aggregate service is instantiated with particular parameters. Ignoring instantiation-specific parameters such as user ID for a while, a service may have exactly the same parameters for all end-users, session-specific parameters, or describe fundamentally different types of the service. In the latter case, exemplified by operating the same service over GPRS and WLAN access technologies, it may be meaningful to specify different variants of a service to keep the information related to a service manageable. The variant selection may depend on access technology or end-user class. Individual variants may also support multiple access technologies and multiple end-user classes. Different variants may be composed of different kinds of service events.

It should be noted that other factors apart from service variants may affect end-user experience of a service. The service quality support may be provisioned differently for different end-users, even though the composition of the actual service in terms of service events would be identical. An example of this, end-user experience of a service can depend on the QoS parameters of the Home Location Register (HLR) profile in 3GPP systems, and hence be different for different end-user classes. Hence, considering end-user experience of service usage, the following holds in the general case:

End-user perceived variant = service variant × service quality support variant.

An aggregate service does not need to have multiple variants associated with it, but can still be instantiated with different parameters.

Parameters include:

- Business oriented parameters
  - Access technology
  - End-user or end-user group
  - Other context-related conditions defining the variant
- Technical parameters
  - Involved service events
  - Means of creating service events, where relevant
  - Supported access technologies
  - Variant-specific service quality policy
- Service quality requirements
  - Service instantiation time
- Service quality characteristics
  - Variant-specific usage
  - Usage pattern.

Above, the means of creating service events refers to functionality which is used to create or authorize creation of new service events. This kind of functionality typically relates to adding new components to session-based services. For example, IMS uses Policy Decision Function (PDF) for authorising new real-time flows within a session (Poikselkä *et al.*, 2004). It is an example of the application of dynamic policies. On the other hand, simple service events such as fetching an Hypertext Transfer Protocol (HTTP) or Wireless Application Protocol (WAP) page are defined as protocol interactions.

Usage patterns are related to the particular variant of the service in question. For example, demographics and geographical coverage are probably different for a GPRS based service than for a WLAN based one, which results in different geographical usage patterns as well as potentially different user basis.

Service event

A service event is a well-defined unit of communication that can be associated with homogeneous event-specific service quality requirements and characteristics. A single service variant can consist of a heterogeneous set of service events. Examples of compositions of end-user services will be provided in Section 6.5. Typically, a service event can be a transaction or a flow, rather than an individual IP packet.

Service events are logically related to a service variant. The composition of a variant in terms of service events may vary, or service events have different default parameters for different variants. Service event specific event creation methods accounted for the above may be used.

Static policies are usually applied on service event type granularity, and will be discussed below. Dynamic policies can be applied where service event creation methods have been specified.

Parameters include:

- Business oriented parameters
  - N/A
- Technical parameters
  - Source and destination IP addresses
  - Port numbers
  - Protocol numbers
  - IPv6: flow label
  - Other classification data
- Service quality requirements
  - Types of requirements: inherent or designed
  - End-to-end delay requirements
  - Delay variation requirements
  - Packet loss requirements
- Service quality characteristics
  - Token bucket parameters
  - Uplink and downlink traffic patterns.

Most of the parameters listed above are technical ones and involve service event or service type specific issues. For example, throughput of Transfer Control Protocol (TCP) is related to end-to-end delay and packet loss characteristics, and the precise functional dependence is determined by the particular variant of the TCP stack in question. More discussions about them can be found in (Poikselkä *et al.*, 2004) and (Räisänen, 2003a).

Uplink and downlink traffic patterns refer to temporal correlations in flows in respective directions, and also between uplink and downlink. For example, Internet browsing consists of small uplink requests, followed quickly by replies which are typically larger in size. Streaming video, on the other hand, consists of control messages that are small in size, and periodic streams of relatively long durations in downlink requirements. Examples of traffic patterns will be encountered later on in this book.

Service event type

Service event type allows for arranging service events into classes which can then be associated with service quality support. For example, traffic from a particular IP address range, traffic relating to a particular user, or services run on a particular protocol can be viewed as service event types for particular purposes. Such an aggregation is useful in allocating service quality support in the domain, which is the case, for example, at the edge of a DiffServ domain. In that particular case, a DiffServ Code Point (DSCP) marking would be associated with a service event type.

Parameters include:

- Business oriented parameters
  - N/A
- Technical parameters
  - Aggregation criterion
  - Service event type specific service quality policy
- Service quality requirements
  - Depends on aggregation criteria
- Service quality characteristics
  - Depends on aggregation criteria.

Due to the definition of service event type as a 'concept of convenience', service quality parameters depend on the chosen aggregation scheme. When heterogeneous service events are aggregated into the same event type, service quality requirements and characteristics may have a broad range. Such aggregation may be needed in systems where the number of simultaneous types or overall available service quality support modes are limited. Interested authors are referred to discussions about multiplexing, treatment of the topic area in (McDysan, 2000).

Policy

Policy is a functionality that spans multiple levels of the entities described above.

Aggregate level, variant level, and service event type can be viewed as having a static policy associated with them. Static policies can be used for providing defaults for certain kinds of entities. The general policies defined on a higher level can be overridden by a lower level, where allowed by the policy framework. In addition to static policies, dynamic policies can be used for controlling instantiation. From the viewpoint of service policies, the most relevant application is the control of creation of service events. The relation of policy to service framework entities is illustrated in Figure 6.2.

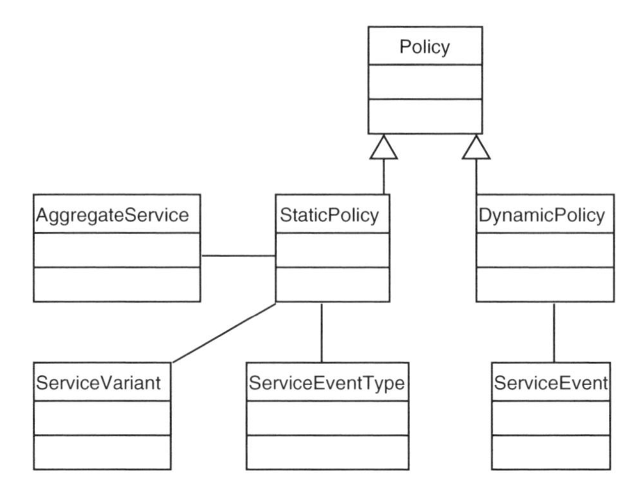

Figure 6.2 An illustration of the application of policy in a service quality framework

In what follows, we shall discuss the use of static policies in managing service quality.

At the highest level is the provider level policy. For example, a service provider may have specific policies defining default values for all services, end-user class specific defaults, access technology specific defaults, or flow-type (event type) specific defaults. On the aggregate level, policies relate to service-level defaults. They may be derived from corporate-level defaults (Koivukoski and Räisänen, 2005).

Service variant-level policies relate to specific needs of particular variants of the services, such as variants for particular access technologies and end-user segments. Variant-level policies relating to 'generic' variant types such as access technology or end-user class specific ones, may be in part derived from corresponding corporate-level policies.

The service event type level policies define the defaults relating to aggregations of service types. They may be derived from corresponding provider level policies. They may be used as a tool in mapping services to resources.

The dynamic service quality policies can be used for achieving more fine-grained control of service events. This is possible since dynamic policy frameworks make use of the 'outsourced policy' mode, where the policy is fetched from a PDF. The rule base of the PDF can be defined to take into account factors such as temporal dependence or element load, for example.

Policies will be described in Chapter 7 as a viewpoint in modelling. Examples will also be provided in that context.

6.2.4 Use of the Framework in Service Quality Management

We shall summarise the use of the service framework in service quality management.

The entities of the service framework can be used for expressing relevant requirements and characteristics relating to service quality. Thus, it is envisioned to be used as part of the service development and implementation, as well as in the handover to operations part of the management framework. The framework can be used in both inter-provider processes and between provider and subscriber. The framework can be used for negotiating service quality, and for mapping it to domain-specific service quality support mechanisms. Examples of the latter will be provided in Part Three.

On the one hand, the design of end-user service usage experience needs to cater for the design of policies, aggregate services, and service events in view of desirable usability. On the other hand, suitable service quality support must exist in a way that is compatible with the desired performance. The framework described above can be used for recording target end-to-end performance levels of service instances, as well as describing service quality support instantiation.

The use of the service quality framework in distributed service environments deserves a special note. The service quality framework described above can be used for describing the requirements and characteristics that individual service events have, viewed end-to-end. In a distributed service environment where the components of a service may be geographically or logically separated from each other, the service framework can be utilised for selecting the optimal composition for a service, where such choices are available. In other words, it is a tool that service composition can use.

The different layers of the service framework can also be related to service management roles, allowing service quality management to also be related to component-based management paradigm (Räisänen *et al.*, 2005; Service Framework, 2004). This mapping also provides a way of relating service quality framework to service life cycle. We shall not discuss this topic here, but it is easy to see that higher-level policies and more generic constructs map naturally to solution-type roles, whereas lower-level policies and constructs are most likely associated with implementation-oriented roles.

Solution oriented service management roles can be associated with the responsibility of defining the aggregate service level and related policies in concert with the product design created by the responsible business role. These kinds of roles are responsible for designing the service that complies with inter-provider agreements. The responsibility for service variants can be associated with technical service design roles. The configuration of rules related to service event types can be allocated to relevant specialists.

6.2.5 Perspective

Comparing the proposed framework to 3GPP QoS model, the proposed framework provides a generalisation for describing multiple subscriber groups or access technology specific variants as a single entity. It also allows for the classification of flows and other kinds of service events for the purpose of service quality support allocation. 3GPP service quality provisioning is a special case of the framework. Similarly, service quality provisioning in DiffServ framework is a special case of the framework.

The framework requires an architecture for actually implementing service quality support. The framework can be used in provisioning service quality in two ways, depending on the type of service quality support:

- Provision static service quality level at the network edge
- Provision allowed parameter range for dynamically instantiated service quality support.

The former case is suitable for classic DiffServ networks, for example, whereas the second option is suitable for dynamic bearer negotiation such as 3GPP framework. The two modes are not mutually exclusive, but can be used for specific needs within a particular architecture. Indeed, this is already possible in the 3GPP R5+ networks.

On an abstract level, the framework is conceptually compatible with the QoS framework reviewed earlier in this chapter (Räisänen, 2004). It should be noted that a practical application of the framework also poses requirements for the provisioning system. For example, in GPRS and 3GPP architectures subscriber QoS profiles are stored in HLR, but APNs require configurations in GPRS Gateway Support Node (GGSN).

Through the use of roles, the entire service management process can be related to the process framework (Service Framework, 2004; SID, 2004). Thus, service management can be related to other processes such as capacity management.

6.3 Security Framework

We shall now move on to the discussion of another important technical viewpoint to service framework, namely, security.

The adoption of Internet Protocol as a convergence layer for services has brought new challenges for ensuring security, the basic reason for this being the fact that the 'classic' construction of IP networks provides little protection against eavesdropping and malicious alteration of traffic. In moving from telephony networks to the Internet, the 'intelligence' in connection set-up has moved from the 'core' (network relay nodes) to the 'edge' (communication endpoints). The routers of the IP domain have traditionally done little more than pass packets to the next-hop router based on the address information stored in packet headers. Indeed, free tools for monitoring IP data transferred in the Local Area Networks (LANs) segments are widely available, for example. Generally speaking, security is most challenging in networks having broadcast segments, such as (non-switched) 802.3 Ethernet or 802.11 WLAN.

6.3.1 Aspects of Security

The basic challenges relating to security relate to confidentiality, integrity, authentication, authorisation, non-repudiation, privacy, trust, and denial of service (DoS). We shall discuss these briefly in the following paragraphs.

Confidentiality refers to the right to keep private communications unavailable to non-related third parties. The privacy of correspondence is a pre-Internet era example of this, demonstrating in its role a basic human right in democratic countries. In addition to the contents of the communication, confidentiality in most cases also covers information about participants to communications.

In the context of security, integrity refers to the prevention of malignant alteration of information. It should be noted that not all alteration is malignant; information may also be altered as a consequence of other reasons such as corruption during transmission.

Authorisation is a process the purpose of which is to make sure that a party seeking to access a resource has the right to do so. Authorisation does not necessarily require the confirmation of an unique identity. Authorisation tokens can be used for authorisation which is not based on identity verification. A key is a familiar example of a repeatedly used authorisation token. An electronic one-time authorisation token is used in IMS for authorising bearers associated with sessions.

Authentication is a process in which measures are taken for ensuring whether a participant to communication has the identity the participant claims to have.

Non-repudiation refers to the means of increasing the probability that a party having carried out a transaction cannot deny having done so. Pre-Internet era examples include signatures in contracts, whereas a 'hash' value added to a message and computed with a personal encryption key is a modern example.

Privacy is often said to mean the right to be left alone. In practice, it means that an end-user or another party should have control of how personal information is used. Location-based Services (LBS) are often used as an example in mobile networks: the end-user must have control over access to personal location information.

Trust is an important concept in modern security. In the context of services, it means the ability to use authentication performed between the end-user and a service for other purposes. Various Single Sign-on (SSO) systems are examples of supporting trust technologies. The federation framework of Liberty alliance seeks to provide a basis for automated trust management.

DoS prevention is usually considered to be a part of security, and amounts to the prevention of normal use of a service. For example, this can be brought about by flooding a server with requests from large numbers of computers infected by a virus. The significance of DoS in the context of security is that systems need to be robust regarding DoS attacks.

As with service quality, we are interested in using the service framework for describing the security requirements and characteristics of a service generically. We shall not discuss domain (technology) specific security issues here. Thus, for example, we shall not discuss encryption, key distribution, authentication, and authorisation methods for wireless access domains here.

Before we move on, a note is in order about an issue that is specific to security, namely, legal aspects. Despite its status as a basic human right, there are cases when society needs to overrun confidentiality and privacy of individuals. Such cases are, or should be, related to law enforcement in democratic countries. Supporting this requires special arrangements and the risk of abuse. We shall not tackle this aspect of security in modelling.

In what follows, we shall next discuss the inter-relations of privacy and trust, and then describe our security framework.

6.3.2 Privacy and Trust

Privacy is of increasing importance for end-users, and relates to issues such as untraceability, unlinkability, and control over personal data (Koivukoski and Räisänen, 2005). The first two relate to the avoidance on the part of the provider to release information which allow third parties to trace the originator of transactions. The third one means that the end-user should be able to control what personal data is used, in particular, operations, and to whom it is made available. The use of temporary identities and pseudonyms is an example of a technology which can be used for protecting privacy.

In the context of modern security, trust refers to the ability of a party to rely on another party. Trust is important in the distribution of cryptographic keys. For example, the end-user may trust another end-user, or a service provider may trust another service provider or identity provider. In Liberty Alliance, the concept of 'circles of trust' is used.

6.3.3 Security Framework

In the following text, we shall describe the security framework for service modelling. We assume the entities shown in Figure 6.1.

We assume that authentication and authorisation for access technology is handled at the network level, and does not need to be taken into account in the security framework. Authentication and authorisation to service is within the scope of this framework. Below, this is referred to as an 'end-to-end' characteristic, differentiating it from a domain-specific one. Authentication and authorisation to access on the one hand and to service on the other need not be completely separated, but the service layer can make use of network authentication and authorisation, for example. This method is used in mobile networks in many cases.

Integrity and confidentiality protection are end-to-end characteristics within the framework. Also, the specification of requirements for domain-specific integrity and confidentiality protection is within the scope. The means for implementing integrity and functionality within domains, on the other hand, are not a part of the current framework. Where a service is implemented in a distributed fashion, integrity and confidentiality protection between components of the service is also a part of the framework.

The security-specific framework concepts can be divided into two groups, namely, service-related security parameters and personal parameters. We shall discuss them in the following text.

Service-specific defaults include the following:

- End-to-end encryption used/not used (Confidentiality and integrity)
- Access technology specific encryption is/is not used
- Authentication and authorisation to service is/is not needed
- Non-repudiation is/is not important
- Privacy parameters
- Trust arrangements (for example federation).

These parameters can be defined on the aggregate service level as defaults, and overridden for specific variants. The implementation of parameters such as access technology specific encryption may vary from domain to domain. Service events or service event types can 'inherit' relevant parts of the parameters and override defaults where relevant and permitted. An example of the latter is specifying that media stream (voice signal) of a Voiceover IP call be encrypted, while other service events remain 'clear-text', as industry speak has it.

Personal preferences include:

- Privacy
- Trust.

Privacy definitions can be end-user–specific defaults, or specific to services or parties. Trust definitions can relate to other end-users or parties.

Referring to the requirements/characteristics classification used for service quality, confidentiality, integrity, and non-repudiation can be viewed as requirements. Privacy could be viewed as being instance-specific requirement as well. The encryption and application of non-repudiation techniques can be viewed as also belonging to characteristics, since they affect the way that a message can be accessed or modified. In Chapter 7, we shall encounter an example of these kinds of side effects of encryption.

6.3.4 Summary

The security framework adds specific issues of interest to the service framework, analogously to the service quality framework. The security parameters can be viewed as being part of the aggregate service definition, and also be related to service variants and service events. Privacy and trust can be viewed as belonging to policies in the form of personal preferences. The requirements and characteristics of a service with respect to security are defined end-to-end, and mapped onto domain-specific security mechanisms along the end-to-end path.

Analogously to service quality, the security aspects of a service may affect the composition of a service in a distributed service environment. We shall not discuss this here.

6.4 Using of Service Framework in Management Framework

We shall next summarise the anticipated use of a service framework within the management framework described in Chapter 5.

As we have seen above, the service framework facilitates the organised representation of technical information related to services. It is thus relevant to processes that specify, use, or exchange this kind of information, either within a stakeholder or between stakeholders. We shall now discuss a few examples of its use in particular situations. For this purpose, the following cases are considered: product creation, service subcontracting, mapping service to resources, purchasing a product, and service optimisation.

6.4.1 Product Creation

Product creation processes define which services belong to a particular product. Target end-user segment for products, and hence also services constituting the product which are visible to end-users, are defined here. Product creation also defines where the product is available, and what kinds of variants exist. Product variants may map to different service variants, where instantiation does not support necessary flexibility. The relevant aspects of the above information are stored in the aggregate service definition. In addition, the service framework can be used for accessing information about existing SLAs towards partners and customers.

Service definition and implementation parts of service life cycle use input provided by product definition to specify detailed service framework parameters by creating necessary service variants, service events, and event types. Suitable tools can be used for reusing the existing service framework entity definitions, either by linking them as parts of new definitions, or by using them as templates in creating new entities.

6.4.2 Service Subcontracting

When subcontracting of services is used, the service framework can be used for information exchange relating to technical requirements such as service quality and security. It is used in defining the subcontracting agreement, and serves as a basis for related service assurance definitions and reporting.

6.4.3 Linking Service to Resources

This set of tasks is valid for new or modified services. The service framework has been used to describe the technical design of the service, and is also used to link service to resources. The actual link may be via other services, or directly to the resources. In the former case, service framework can be used for relating to the technical parameters of the 'mediating' service. In the latter case, the service framework is used for determining that resource provides adequate support for service quality and security. For security and service quality, this involves checking that necessary mechanisms are supported by the resource, and that sufficient amount of service instances can be supported.

6.4.4 Purchasing a Product

In selling a product to a customer, the service framework is used to describe a view to planned performance to the product, and used as a tool in possible negotiations. It also forms the basis of SLA definitions and reporting.

6.4.5 Service Optimisation

Service optimisation uses service framework to fetch information about planned performance level and to compare it to actual performance levels. Potential suggested changes pertaining to the composition of service resulting from service optimisation, as well as potential changes to target performance levels of individual services are recorded using the service framework.

6.5 End-user Services

We shall now provide an overview of requirements and characteristics of selected classes of end-user services. The purpose is to provide an insight into what typical end-user services look like from the viewpoint of service quality. We shall make references to the service quality and security framework described above. This overview is meant to provide a more concrete basis to the concepts discussed above. The discussion is based on previous work in (Laiho and Acker, 2005) and (Koivukoski and Räisänen, 2005). We shall use familiar classes of services in the interest of clarity, and also for the reason that almost all current and near-term services can be composed of service events belonging to the classes discussed below. We shall not consider broadcast IP services such as Digital Video Broadcasting for Handhelds (DVB-H) here.

The discussion below will concentrate on the technical requirements and characteristics of end-user services with reference to the service events they are composed of. More complete use cases will be provided in Part Three.

End-user services are classified as follows in this section:

- Background data transfer
- Interactive data transfer

- Messaging
- Streaming
- Conferencing.

End-user services can be classified in different ways. The above classification has been done from the viewpoint of service quality. Comparing the above classification to 3GPP traffic classes, basically messaging has been added to the four traffic classes of the 3GPP QoS framework. We shall discuss justification for this below. Even though the names of the end-user service classes used here are closely related to 3GPP traffic classes, it is important to note that as such, the latter are conceptually different from end-user services. An end-user service may involve multiple 3GPP traffic classes simultaneously.

We shall discuss the classes below one by one. Prior to discussing end-user service class-specific issues, we shall start with a summary of some generic issues.

6.5.1 Introduction

As discussed previously, end-user services and their constituents have service quality requirements and characteristics which either follow inherently from the nature of the service events, or are designed by providers, depending on the service in question.

The descriptions given below for end-user services are phenomenological, and do not use numerical values. The intention is to provide an overall comparison of the requirements and characteristics of constituents of end-user services. Quantitative values for some services are discussed in, for example, (ITU-T Recommendation G.1010, 2001) and (Räisänen, 2003a).

The transport protocol (Layer 4, L4 in ISO/OSI classification) used for delivering packet-based content potentially affects end-to-end service quality. Some Layer 4 protocols such as TCP support retransmission of data on L4, which potentially affects delay and packet loss experienced by flows during transmission. Similarly, link layer (L2) technologies such as fragmentation and scheduling algorithms may have an impact on end-to-end service quality. We shall not discuss these further here, and an interested reader is referred to (Armitage, 2000; Halonen *et al.*, 2003; McDysan, 2000), or (Räisänen, 2003a).

Due to the fact that certain aspects of security such as confidentiality and integrity basically permeate all end-user services, they are not listed separately below. They should be read as implicitly belonging to the list of requirements given below.

6.5.2 Background Data Transfer

Background data transfer refers to the exchange of information that does not involve the end-user interactively, even though the end-user may have initiated the service, either directly or indirectly. The background transmission of e-mail messages is an example of this class, and the background downloading of music from the Internet is another. Background data transfer may have aspects that are visible to the end-user, especially when the amount of data to be transferred is large. In such a case, it is preferable that the overall duration of downloading is predictable.

The background data transfer typically consists of one service event, which is the data transfer flow. No session is associated with this class of service, even though the transfer may be initiated from a session. The requirements for background data transfer end-user service class include:

- Reliable delivery end-to-end
- Stable throughput preferred for large service events.

Characteristics for background data transfer include:

- Often predominantly unidirectional
- Content size maybe large.

In addition to downloading, the background data transfer can also relate to uplink transfer, such as backup of data and/or configuration stored in communications endpoint.

From the service quality viewpoint, background data transfer is the simplest and least demanding service.

6.5.3 Interactive Data Transfer

The interactive data class of end-user services is based on real-time interaction with the user. Consequently, responsiveness is important for this end-user class. Typically, interactive traffic is based on request/reply pattern using HTTP, for example. Depending on the service in question, the size of the reply event may be small or large. For example, an acknowledgement of a transaction requires only a small message, whereas a response to the request for a picture would most likely be relatively large. Involved service event types include request message and reply. The latter can consist of a flow of multiple packets. Interactive data transfer may or may not be associated with a session.

Typical requirements include:

- Service instantiation time relatively short
- Low packet loss for uplink and downlink preferable
- Stable packet throughput preferable for reply events containing large amounts of data
- It should be possible to protect the privacy of end-users.

Typical characteristics include:

- Uni-directional or bi-directional
- Request size typically small
- Size of reply may be small or large
- Request and reply temporally correlated.

Comparing interactive data transfer, two observations can be made. Interactivity leads to the inclusion of service instantiation time as a service quality parameter of interest. Interactivity also leads to indirect effects; for example, the requirement for low packet loss contributes to higher data throughput (Padhye *et al.*, 2000).

Secondly, the interactive data transfer consists of correlated pairs of uplink and downlink traffic. This leads to patterns in service characteristics.

6.5.4 Messaging

Messaging is a generic class forming the basis for services such as e-mail, chat, group chat, and push-to-talk. The most important delineation from interactive transfer is that messaging takes place between end-users, even though the messages may be handled by intermediate servers. Messaging can be stateless, or associated with a session. The degree of interactivity may vary from 'background-like' (e-mail) to 'interactive' (chat/push-to-talk). Involved service event types include: session management; uplink and downlink message delivery.

Typical requirements include:

- High availability
- Service instantiation time relatively short
- Reliable delivery of messages
- End-to-end delay can be relatively large for non-interactive messaging but should be in the 'interactive' category for services such as chat and push-to-talk
- It should be possible to protect the privacy of end-users. Support for pseudonymity may be desirable for chat.

Typical characteristics include:

- Typically bi-directional
- Content size varies
- Interactive messaging: temporal correlation for service events within a session
- E-mail synchronisation: a number of service events are transmitted in a temporally correlated manner.

As an end-user service, messaging shares many of the characteristics with interactive data transfer and background data transfer. There are certain reasons for depicting it as a separate category, however. The first one is the importance of messaging type services. Services such as e-mail are crucial for the operation of developed societies. The second reason is the fact that operation of messaging service may require configurations which are different from the ones for accessing information in servers.

Messaging makes use of interactive data transfer and background data transfer, but requires something else on top of it. It can be associated with policies for inter-person communications, which may be different from the ones devised for access of server-based information. End-users tend to have a different disposition towards inter-person services than ones involving communications between a human user and a machine.

From the viewpoint of security, messaging can also be associated with a set of service-specific requirements which are typically more clearly delineated than ones for interactive data transfer in general. The implementation of messaging on a dedicated server leads to specific service quality related requirements for the server, which may be different from the generic requirements for interactive data transfer.

6.5.5 Streaming

Streaming refers to the transmission of a continuous media stream across the network. The most obvious examples of media involved may include audio and video, but also telemetry data and control for critical remote control applications can belong to this class.

Characteristics of media stream vary according to the content and the encoding applied: the speech signal may consist of a string of talk spurts and silence periods, whereas music and video are typically transmitted using continuous streams. For video, the momentary bit rate of the stream can be very variable. Streaming is typically session-based, and media streaming is associated with controls such as play, pause, rewind, and fast forward. The service event types involved include control signalling and one or more media streams.

Typical requirements include:

- Service instantiation time short
- Media streams: packet loss can be tolerated
- Control signalling: interactive responsiveness
- End-to-end latency can be relatively large
- Delay variation can be relatively large
- Stable throughput required for media stream
- It should be possible to protect the privacy of end-users.

Typical characteristics include:

- Control signalling uplink, media stream downlink
- Control signalling temporally irregular
- Media stream continuous on the average
- Video stream may appear in bursts if it is not shaped.

As a service, the special feature of streaming is the combination of continuous media stream with interactive class media control streams.

6.5.6 Conferencing

Conferencing class services are considered to have voice telephony at a minimum, and may have a video component as well. These real-time media components may be supplemented by group work support functions such as white-boarding and chat.

Typical requirements include:

- Availability is high
- Reliable transport and relatively low delay for control signalling
- Media streams: low delay, stable throughput, packet loss can be tolerated, but should not be temporally correlated
- Session control signalling for supplementary session components has interactive responsiveness.

Typical characteristics include:

- Service event invocation pattern random within sessions
- Media stream is typically bi-directional.

Conferencing is the most demanding service in multiple respects. It can be compared to legacy services such as Public Switched Telephone Network (PSTN), and is expected to have comparable support from the business criticality viewpoint. The service quality requirements for the telephony media stream are the strictest of all services, and the requirements for telephony control signalling are also demanding. As such, carrier-grade telephony can be said to be a veritable yardstick for service quality in packet-based networks. When additional components such as white-boarding are present, requirements for control signalling related to them is 'merely' interactive.

6.6 Summary

We have presented a framework for describing requirements and characteristics with respect to service quality and security. The framework can be used by service management processes relating to products.

Most important classes of end-user services were analysed in terms of requirements and characteristics to gain insight into the composition of typical packet-based services. Comparing conferencing to streaming, for example, the former is bi-directional and has stricter requirements regarding service quality. Both conferencing and streaming are session-based, whereas simple services such as downloading can consist of individual service events. In the most complex case, the service event types making up a service include session management signalling, media flows, and possible other session components.

Some service events such as media streams of Voice over Internet Protocol (VoIP) telephony have inherent service quality requirements associated with them, whereas for some services such as data downloading, service quality can be engineered. The framework needs to accommodate both types.

Session-based services are typically more demanding regarding availability and continuity of sessions, since they involve multiple end-users simultaneously. In general, for all interaction sequences, the feeling of responsiveness is important. The introduction of real-time service events such as the transmission of voice or video brings with it the need to have resource effectively put aside for them. Finally, the importance of predictable service experience cannot be over-stressed.

The service framework is a construct that can be used in product development as well as customer-oriented, service-related, resource-related and partner-related process areas of the management framework. The service framework serves as a tool for designing, agreeing upon, supervising, and reporting service quality for services. End-user experience of service quality is affected by the service quality support of the connectivity provider, in addition to the service quality of actual end-user services. Furthermore, as we shall see later on, the service quality related definitions can be attached to services which are not directly visible to end-users.

6.7 Highlights

Ten things to remember from this chapter:

- The service framework is described, covering service quality and security.
- The adoption of IP as a basis for service convergence has necessitated focusing on service quality and security.
- The ITU-T differentiates between the QoS requirements of customer, the QoS planned by the provider, the QoS delivered by the provider, and the QoS perceived by an end-user.
- The 3GPP QoS model includes four traffic classes and a set of QoS attributes.
- We shall use the following concepts: aggregate service, service variant, service event, and service event type.
- The end-user–perceived service quality is affected by service variant, service quality support variant, and instantiation parameters.
- The policies can be used for structuring management information on multiple levels.
- The security concerns include confidentiality, integrity, authentication, authorisation, non-repudiation, privacy, trust, and DoS.
- Privacy is of increasing importance as services grow in complexity.
- End-user services can be classified into background data transfer, interactive data transfer, messaging, streaming, and conferencing.

7

Service Modelling Patterns

As we have learnt earlier, service modelling is an enabler for structuring product and service creation, as well as for other parts of product and service life cycles. The service model needs to be amenable for use by the management framework, and includes the service framework described in the previous chapter as one building block.

In what follows, views to a service model are described in terms of modelling framework and patterns. The pattern-based approach was chosen in order to facilitate both 'ontology' type models common to multiple domains, as well as domain-specific, limited-purpose models. The purpose is that relevant service modelling patterns can be used for both types of models. Furthermore, a pattern-based approach allows us to cover aspects of a relatively wide technology area.

We shall start with a description of the modelling framework, and proceed to describe some of the relevant patterns. We shall discuss how the patterns can be used in a specific service model, and discuss their relationship to existing models. We shall conclude this chapter with a summary.

Examples of the use of some of the patterns will be provided in Part Three using three cases.

7.1 Modelling Framework

The *raison d'être* of service modelling is providing information modelling to serve as a basis for the service management processes. This means the ability to represent linkages and dependencies among products, services, and resources. The actual service model may vary from environment to environment, depending on the way that products, services, and resources relate to one another. The basic structure of the inter-relations between central concepts, however, often exhibit the same structure in different environments. The description of these relatively invariant parts is the topic of this chapter. The role of the modelling concepts discussed in this chapter is illustrated in Figure 7.1. Different environments may implement different parts of the entire set of service modelling patterns. We shall mention a few examples of 'packaging' at the end of the current chapter.

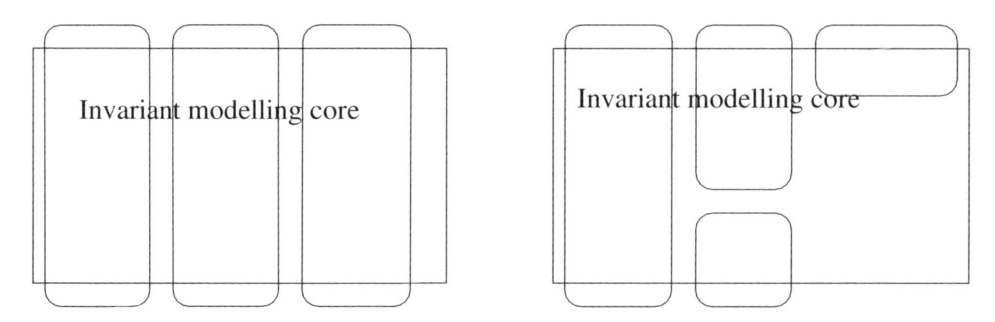

Figure 7.1 An illustration of the role of invariant modelling concepts

In the following text, we shall present the modelling framework that is used in this book. It defines the 'vocabulary' for service modelling, and shows the inter-dependence of entities on an abstract level. We shall not describe the actual schemata here, even though the Unified Modelling Language (UML) diagrams drawn with ArgoUML can be exported to XML Metadata Interchange (XMI) format.

It is assumed that the modelling framework can be used – for relevant parts, of course, by different stakeholders of the service management process. Typical examples of the use of the framework include the following:

- The stakeholder responsible for the product uses the framework for constructing a description of the characteristics of the product as well as the relevant aspects of the services involved
- The stakeholder responsible for aggregating a service uses the framework for constructing a description of the characteristics of the aggregate service as well as the constituent services
- The stakeholder responsible for providing elementary services uses the framework for constructing a description of the characteristics of elementary services
- The stakeholder responsible for the operational management of services or resources uses constructs based on the framework for representing performance data in the context of services
- The service developer uses constructs based on the framework for implementing a component for a service
- The role responsible for activating the service uses constructs based on the framework as a reference in specifying parameters for service components
- The end-user uses a model derived from the framework for expressing service quality and privacy preferences.

We shall not describe the division of service functionalities into components here. The division depends on the details of the technological environment and also potentially on the value net in question.

We shall discuss the application of the model towards the end of the current chapter in section 7.3.

7.2 Modelling Patterns

In this section, we shall present service modelling patterns using UML notation. They build on the preceding discussions, and are based on studying inputs such as the Telemanagement Forum (TMF) Shared Information Data (SID) model and related efforts in EU MobiLife project, to name the major sources of input. The patterns can be viewed as the illustrations of generic inter-relations of concepts. The set of patterns provided here alone is not sufficient to form a service model.

The patterns provide building blocks of concepts to be used in individual service models. The patterns displayed are to be considered to be examples of relations to be taken into account in devising actual service models, rather than seeking to be consistent ontologies. The reader is referred to models such as TMF SID, Enhanced Telecom Operations Map (eTOM), and New Generation Operations Support Systems (NGOSS) for the latter purpose. In the interest of clarity, relations are shown with views containing a small number of entities. The patterns have been generated as views from a single UML model, which provide a certain degree of consistency.

Notes about the usage of the patterns in a model describing a specific environment will be made in section 7.3. At the moment, suffice it to mention that no differentiation is made between abstract and concrete classes. This separation needs to be done when the coherent model is put together. Furthermore, generally speaking, meta-modelling and instantiation-related issues are not considered here, but patterns are provided to present a set of relations using the UML format.

The patterns have been classified into broad categories in the interest of clarity and ease of access. The categories include abstract patterns, basic entities, and miscellaneous patterns. Abstract patterns can be used in multiple contexts, and represent modelling at the most abstract level. Basic entities describe the entities which are specific to service modelling. The miscellaneous patterns subsection describes patterns which are frequently encountered during service modelling in different environments.

Regarding the 'basic entities' and 'miscellaneous patterns' categories, it is useful to point out that descriptions of the patterns build on the requirements and the preceding discussion found in this part of the book. Also, the descriptions of patterns make references to other patterns. Thus, the pattern descriptions of the last two types can be considered to be specific views into service model.

In what follows, we shall describe the individual patterns by means of a textual description and UML diagrams. Next to the UML diagrams, the entities depicted are described. We shall not describe the multiplicities for entity inter-relations, and shall also not name the inter-relations. This choice was made to shorten the descriptions of the individual patterns, and, therefore to accommodate more patterns.

7.2.1 Abstract Patterns

We shall start our review of patterns with abstract patterns. In this book, they are perhaps the closest relatives to classical software UML patterns (Gamma *et al.*, 2004). As such, they can be applied in a variety of circumstances.

Role

A particular entity may have different meanings for different stakeholders. The concept of role allows for the representation of stakeholder specific 'interfaces' to entities. Generally speaking, it can be used to represent interfaces to a particular entity from a specific point of view. Thus, a single entity may play simultaneously multiple roles from different viewpoints. The role pattern is used in TMF's SID model, for example.

A role can represent a set of activities, and can be used to identify relevant process areas. These types of use of roles were previously discussed in the context of the TMF Service Framework, and a few examples of service management roles were provided in section 5.2.

Below are three examples to illustrate how roles can be used in modelling.

The first example is not serious, and is intended to get us started. Let us consider a model for a dog and its owner. They can both be described 'objectively', by listing attributes related to each of them. A dog, for example, would have attributes relating to its colouring, the length of its hair, and so on. Such an 'objective' description is devised by a human, and as such, encompasses one viewpoint only.

To enrich this 'system description', we can use roles for encapsulating the stakeholders' viewpoints of each other. Thus, we could have:

Dog (owner's view):

- Has been trained to obey orders in most cases
- Has been trained to take a walk twice a day
- Is cute.

Owner (dog's view):

- Has been laboriously trained to feed me relatively regularly
- Has been trained to take a walk twice a day
- Spends too much time in front of the bright rectangle.

The owner views a leash as a means of controlling a dog's behaviour during the daily outdoor endeavours, whereas the dog might view the leash as being a way to keeping the owner happy and increasing the odds of eliciting a tasty snack after the outing.

The second, more serious, example (Figure 7.2) shows how a leased line can be modelled by a provider and a customer. Both stakeholders perceive the same entity through different roles. From the viewpoint of the connectivity provider, the leased line is a service provided to customers. From the viewpoint of the subscriber, on the other hand, the leased line is both a service (subscribed to) and a resource (having the specific capacity to be used by applications).

Description of the entities in Figure 7.2:

- Connection: a generic class for representing connections
- LeasedLine: a specific type of Connection representing leased lines
- ResourceRole: a resource-oriented view to LeasedLine
- ServiceRole: a service-oriented view to LeasedLine

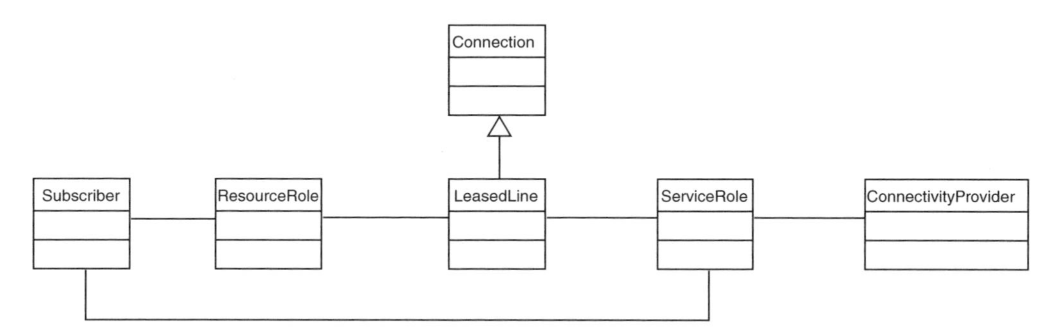

Figure 7.2 Role example 1: use of roles for a leased line

- Subscriber: as described earlier
- ConnectivityProvider: as described earlier.

The third example shows two different process areas interfacing to a network element – the edge router – via different roles. The example involves two kinds of service management roles, namely, an implementer role configuring charging rules to an edge router, and a connectivity builder role configuring traffic conditioning rules. The two service management roles see different aspects of a single device, whereby it is natural to use roles for the representation of the relevant aspects of the resource for both.

Description of the entities in Figure 7.3:

- EdgeRouter: a Resource (inheritance not shown)
- ChargingRole: the charging-oriented view on EdgeRouter
- TCRole: the TC oriented view on EdgeRouter
- Implementer: a set of tasks relating to configuring charging
- ConnectivityBuilder: a set of tasks relating to configuring the connectivity parameters, including TC.

Obviously, it would be easy to construct many more examples of the use of roles. For the rest of the book, we shall mostly use the two approaches described above, i.e., the representation of different views to an entity and the description of a set of tasks related to an entity.

Aggregation

As described in Section 4.1, aggregation is a part of the UML notation. We shall complement the preceding description by adding some general comments about the use of aggregation in modelling. As with other abstract patterns represented here, the considerations are meant to be applicable to a range of models.

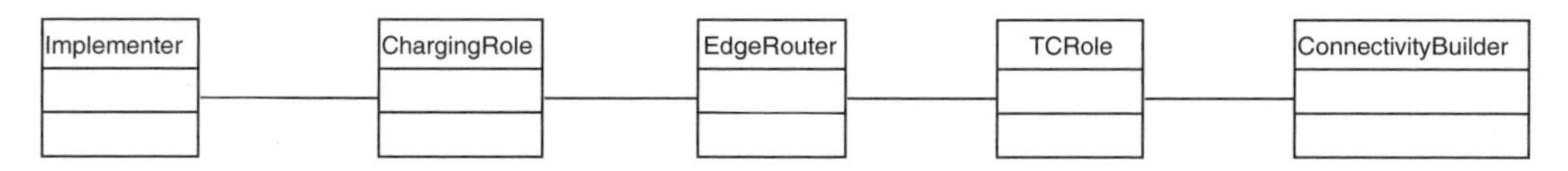

Figure 7.3 Role example 2: use of roles for interfacing to a resource

The basic purpose of aggregation is to indicate that an aggregate entity consists of other entities. Some of the constituent entities may be mandatory, whereas others may be optional. In UML, optionality can be expressed with multiplicity starting with zero, for example, `0..N`, whereas mandatory dependence would be expressed as `1..N` where at least one constituent would be necessary.

Regarding mandatory constituent entities, there are special cases relating to the representation of back-up resources and clusters. These will be described later on.

When using aggregation, care should be taken to avoid undesirable effects. Collecting heterogeneous constituent entities together may lead to conflicts or additional complexities on the aggregate level. To avoid this, it may be useful to aggregate entities of the same type prior to performing higher-level aggregation. The documentation of TMF SID model has examples of this form of avoidance of perils.

Containment as a special case of aggregation is supported by UML. We shall not use containment in this book.

Generally speaking, aggregation can be implemented statically or dynamically. In the former case, mapping of constituents to aggregate the entity is determined as part of the definition of the instantiation of the model. Static aggregation does not need special functionalities in order to function. Dynamic aggregation, on the other hand, needs the means of discovering the constituents and performing dynamic binding to them. In a complete model for the system, discovery and binding means need to be listed as resources in the system model.

Resilience

Resilience is a pattern which represents the capability to dynamically replace an entity instance with another in case of failure of the former. Familiar examples of resilience include multiple parallel computers on board a spacecraft. More down to Earth, the reasons for equipment failure may be less exotic than gamma radiation, but the conclusions are the same in mission-critical environments. Resilience is the bread and butter of carrier-grade systems, and service modelling needs to support it.

In terms of modelling, resilience on the basic level amounts to the capability of representing a system of multiple entities such that the order in which they are used in case of failure can be predefined. An example of this is shown in Figure 7.4. Please note that this pattern can be applied recursively, so that also tertiary back-up systems, for example, can be represented as resilience pattern for secondary back-up.

Description of the entities in Figure 7.4:

- Capability: an entity of unspecified type
- PrimarySupport: primary support for Capability
- FailoverSupport: secondary support for Capability, invoked in the event of a failure.

In principle, resilience could be described in a resource-specific manner. Here, we have opted to use a generic pattern for different situations requiring resilience. The resilience pattern described above represents the fact that a back-up entity needs to exist for a primary entity. The identity of the entities in question is resolved during instantiation.

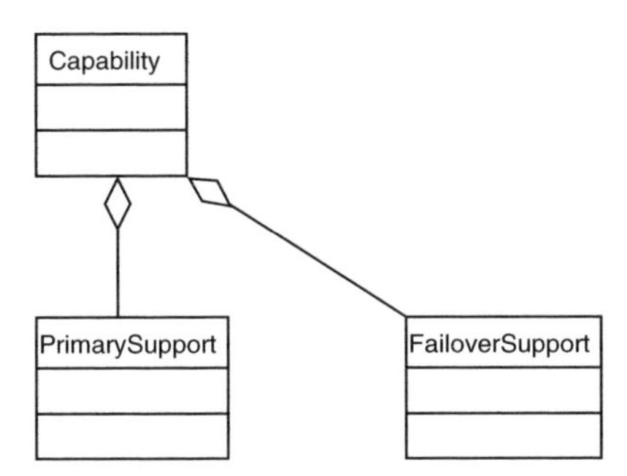

Figure 7.4 Resilience example: representation of back-up capability as a pattern

In practice, resilience can be implemented in different ways. The back-up resource may be inactive up to the point of failure. Another possibility is that the tasks of the failed element are offloaded to an element that is previously active. In the latter case, it is obviously important to be able to assess the capacity implications of the failure. We shall not address this in modelling.

At a given point in time, it is typically important to be aware of whether resilience capability is available or not. For the stand-by type back-up functionality, the operational condition of the secondary resource needs to be ascertained and reflected in the run-time resource inventory. When the tasks of an operational element A are offloaded to another element B in use, it is also important to know whether there is a back-up arrangement available should element B fail.

The resilience pattern can be applied to a diverse set of environments. It can be used for modelling hardware failover arrangement as well as resilience on a service level.

Cluster

Clusters are commonly used for high-volume transactional processing. A cluster consists of a number of elements capable of performing the same set of tasks. A cluster is able to perform automatic load balancing by distributing incoming requests to members of the cluster. This can be realised by having a cluster head for making load sharing decisions, or by performing load sharing in a distributed manner.

In the former case, the cluster consists of a cluster head and a number of cluster members, as depicted in Figure 7.5. Since the cluster head plays a pivotal role in the functioning of the cluster, resilience needs to be implemented. Here, the resilience pattern of Figure 7.4 is used.

Description of the entities in Figure 7.5:

- Cluster: cluster of entities
- ClusterHead: head of the cluster
- ClusterMember: member of the cluster
- PrimaryCH: Primary cluster head
- SecondaryCH: secondary cluster head, invoked in case of failure.

The failure of a cluster member affects the load of other elements indirectly. When the cluster head is present, it needs to be aware of the operational cluster members.

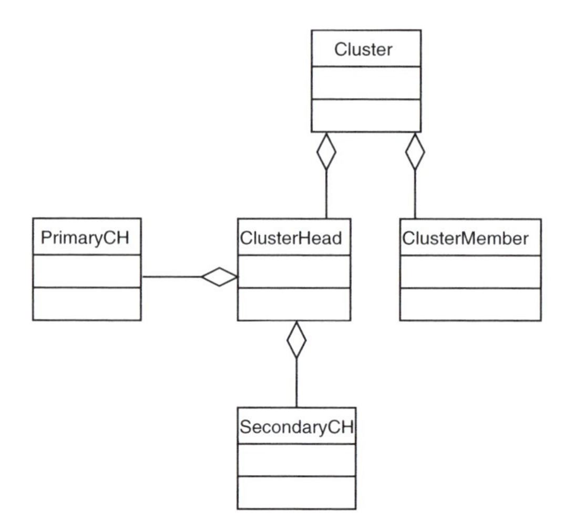

Figure 7.5 Clustering example with cluster head

In dynamic clusters, the cluster head needs to be resolved dynamically. Analogous to determining the effect of the failure of an element belonging to a resilient set-up, this needs help from a real-time resource inventory.

Template

A template is a pattern representing the re-usability of information through templates containing parts that are common to multiple entities. In the TMF SID model, the specification pattern corresponds roughly to what is called a *template* here. In an operational environment, a template can define the set of parameters that are common to multiple devices. From the procedural point of view, a template is a tool that can be used for increasing coherence.

The reason for representing templates as a separate pattern is that different service management roles may be responsible for creating and managing a template than for using the templates. In (Service Framework, 2004), for example, separate service management roles are responsible for specifying services and implementing them. These roles can be viewed as utilising templates by creating and filling in templates respectively. Templates and entities derived from templates may also be hosted on different platforms, as shown in Figure 7.6. In the example, templates and derived entities are hosted on two separate platforms, and three separate roles are responsible for the template and the entities.

Description of the entities in Figure 7.6:

- Template: a template for an entity
- Platform: the platform which relates to a Template
- SpecRole: the service management role responsible for specifying a Template
- DerivedEntity1, DerivedEntity2: the entities created using a Template
- OperRole1, OperRole2: the roles responsible for operating and/or specifying DerivedEntities
- Platform2: the platform relating to DerivedEntities.

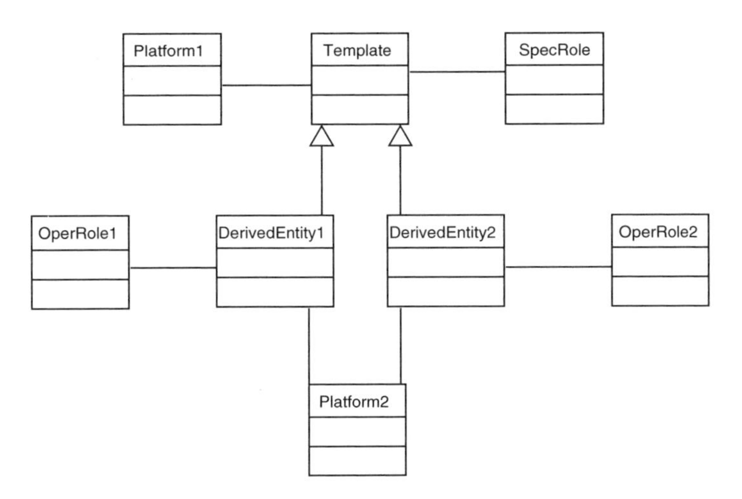

Figure 7.6 Template example

Above, the use of templates in connection with services has been discussed. Templates can, however, also be used for specifying other kinds of entities such as products and resources.

Metadata

Earlier, we reviewed some reasons for adding metadata to services, including faciliation of automated service composition. Utility of metadata is not limited into dynamically composed services, but also has its used for statically composed services as well. Being able to automatically check the validity of a service in a broader context helps in avoiding mistakes.

In general, metadata can be associated with any entity for which automated operations such as matching need to be performed. Metadata related to this end helps in identifying the object more precisely for a specific purpose. Modern image processing software is able to associate information about the geographic location, the time, the date, as well as user-supplied information with a digital photograph, allowing searches such as 'all photographs taken in Helsinki, Finland, in July 2005' to be matched in a search.

Metadata can also be used in limiting the way in which an object is used, which was the case in (Jones, 2005). The example given in the article was related to the accelerator pedal of a motor car, with metadata specifying the allowable angular range of the pedal. In the near term, this type of metadata can be used as a tool for avoiding the erroneous use of entities. In the long run, such data can be expected to be essential for dynamic composition schemes such as Semantic Web.

Metadata requires that an ontology – specific-purpose or common – be available to the creator of metadata as well as to users. In some cases, a specific ontology can be created on-the-fly using reasoning methods.

Basic concepts related to metadata are shown in Figure 7.7.

Description of the entities in Figure 7.7:

- Entity: an entity
- Metadata: the metadata relating to Entity

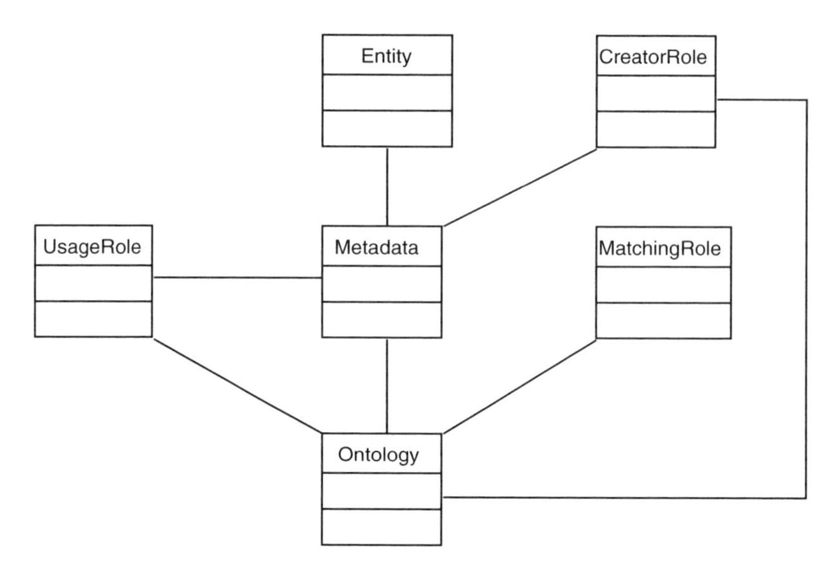

Figure 7.7 Metadata relations

- CreatorRole: the role (human) responsible for creating Metadata
- UsageRole: the role responsible for utilising Metadata
- Ontology: the ontology used by Metadata
- MatchingRole: the role responsible for assessing the match based on Metadata.

7.2.2 Basic Entities

We shall now move on to describe the basic entities of a service model. Below, we shall describe agreement, resource, customer, policy, product, service, configuration, session, and service event. Roughly speaking, the first seven have counterparts in the TMF SID model, whereas the last two have been added to cater to the service quality and security aspects of the service frameworks discussed above.

Agreement

An agreement is a formal record of the conditions to be applied to a business relation between two or more parties. In the inter-stakeholder classification that we are using, we assume that an agreement relates to the products purchased by a stakeholder from another stakeholder.

An agreement may include generic, business-oriented, and technical parts, which are discussed below. In addition to listing details of the contract, an service level agreement (SLA) also serves other purposes, such as being a communications medium. As illustrated in Section 5.2, an agreement may be used in many different inter-stakeholder relations.

The generic part of the agreement describes the scope of the agreement, including information about involved parties, products, and the definition of the terms of applicability. The possible specification of the temporal duration of the agreement and the definition of conditions constituting normal and abnormal situations in view of the agreement belong to parts of the terms of applicability. The generic part may also describe possible procedures to be followed in abnormal situations, and in revising the agreement, if need be.

The business part of the agreement describes monetary transactions such as remunerations and other kinds of compensations. The magnitudes of transactions related to abnormal conditions such as failure of a party to follow the agreement shall be specified here where relevant.

The technical part of the agreement describes the necessary technical parameters, as well as related supervision and reporting procedures.

As a specific example, we shall describe the parts of an SLA in a DiffServ domain using the terminology of (Grossman, 2002) referred to in section 3.9.1. An SLA is a special type of an agreement for specifying the service level between a customer and a provider. The SLA includes other parts, including the TCA specifying traffic classification rules to be applied in the provider/customer interface, SLS describing the technical parameters in terms of the measurable service quality characteristics, and TCS enumerating the parameters used in conditioning. The SLS component may make reference to domain-wide characteristics by using Per-Domain Behaviour (PDB) concepts.

Figure 7.8 shows a simple model for the inter-relations of DiffServ-related SLA entities. DiffServ SLA is a special case of a more general SLA, and aggregates DiffServ SLS and DiffServ TCA. DiffServ TCA, in turn, aggregates DiffServ TCS. Finally, DiffServ SLS uses DiffServ PDB.

Description of the entities in Figure 7.8:

- Agreement: used previously
- SLA: SLA, a type of agreement
- DiffServSLA: DiffServ SLA, a type of SLA
- DiffServTCA: DiffServ TCA, a part of DiffServSLA
- DiffServTCS: DiffServ TCS, a part of DiffServTCA
- DiffServSLS: DiffServ SLS, a part of DiffServSLA
- DiffServPDB: DiffServ PDB, used by DiffServSLS.

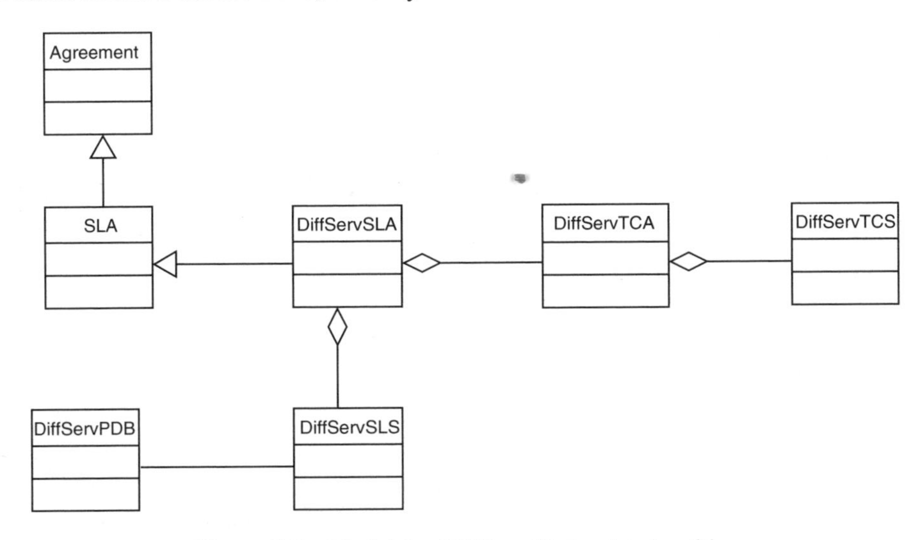

Figure 7.8 Model for DiffServ SLA related entities

Not depicted in Figure 7.8, DiffServ SLA and DiffServTCA can be classified as business-oriented entities, whereas DiffServTCS, DiffServSLS, and DiffServPDB may be viewed as technology-oriented entities. This has a bearing on role mapping. Please note that although TCA has been shown to be a part of SLA, this is not always the case.

As discussed earlier, the agreement may either be explicit or implicit in nature. The degree of exactitude of the agreement may vary greatly. An implicit agreement between two parties of peer-to-peer services is likely to be less specific than an agreement between an Internet access provider and a corporation.

Agreements, or parts of them, can be constructed in an automated manner. This requires a suitable framework to be in place in order for this to work. In the envisioned future, broker-based systems are expected to participate in this. More near-term and common examples of automation of agreements are situations in which the use of a service implies an agreement with the provider.

Resource

Resources represent the technical functionalities that are needed by services. Services tap into a resource, controlling the sharing of the capacity of the resource to perform a specific task. Resources may be tangible or intangible in nature. Specific computer hardware is an example of the former class, whereas an Apache web server software running on a computer belongs to the second class. By and large, logical resource can be software such as an application program or an operation system. In TMF SID model, the two types of resources are called *physical and logical resources* respectively, and we shall also use this terminology here. Logical and physical resources may both require maintenance and configuration and can be composite entities of the respective type.

The web server example calls our attention to the relationship between logical and physical resources. Obviously, logical resources require physical resources to operate. A single physical resource may support multiple logical resources at the same time. In a multi-processor environment, a logical resource may not be tied to a particular physical resource, but may execute on multiple instances of a physical resource. From the view-point of modelling, both cases can be covered with a relation in which a logical resource needs a physical resource of a specific type to operate.

A resource is typically associated with a set of maintenance procedures, which can be associated with a role. Similarly, a resource typically has a configuration associated with it. Furthermore, the configuration can also be associated with a role that is not necessarily the same as the resource maintenance role. The above considerations bring us to the basic model for the resources shown in Figure 7.9.

Description of the entities in Figure 7.9:

- Resource: a resource
- LogicalResource: a resource which is abstract
- PhysicalResource: a resource which is physical.

The abstract patterns described earlier in this chapter can be used for modelling load sharing and resilience of the resources. In the most general case, resilience and load sharing need to be considered for logical and physical resources separately.

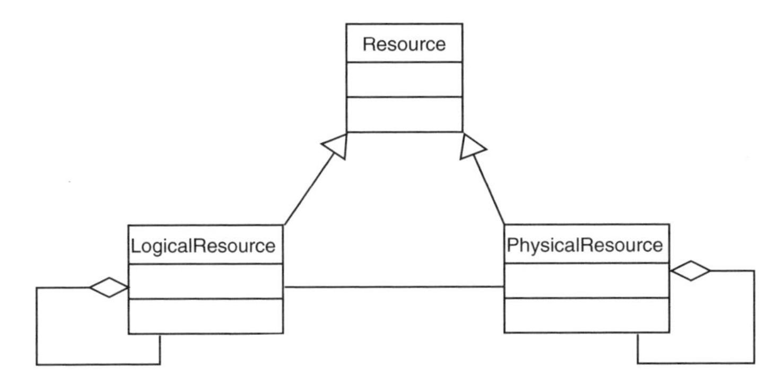

Figure 7.9 The basic relationship between logical and physical resources

Customer

The basic facts relating to the customer were discussed in section 5.2, and we shall make some complementary comments here. We shall not consider the marketing-oriented aspects of customers in this book, and an interested reader is referred to TMF SID, for example. We shall also not attempt to provide a complete set of customer modelling patterns here.

An aggregate customer such as a corporation – or a part thereof – can have multiple end-users and make agreements on their behalf. There can either be separate agreements towards different providers, or an agreement with one of the parties involved implies agreements towards others as well. For example, end-users can have direct relations to different parties involved, related to preferences.

Armed with the concept of roles, we can add some further details to the modelling of customers, in addition to the ones represented in Figure 5.6. An end-user can be a direct subscriber to the services for which the aggregate customer does not have agreements. This can be represented with the customer role which the end-user plays in these situations. The situation can be modelled with aggregate subscriber and customer role being the specialisations of a generic subscriber class as shown in Figure 7.10. The figure also depicts two forms of agreements, namely, explicit and implicit ones.

Description of the entities in Figure 7.10:

- Subscriber: used previously
- Agreement: used previously
- AggregateSubscriber: aggregate subscriber multiple End-Users
- End-User: used previously
- SubscriberRole: The end-user uses this role to interface when subscribing to services
- ExplicitAgreement: an agreement formed explicitly
- ImplicitAgreement: an agreement not formed explicitly.

The agreement between the subscriber and the provider relates to one or more products, specifying which services are in scope of the agreement. Above, we have only discussed the application of the role of the customer to the subscriber and the end-user. Referring

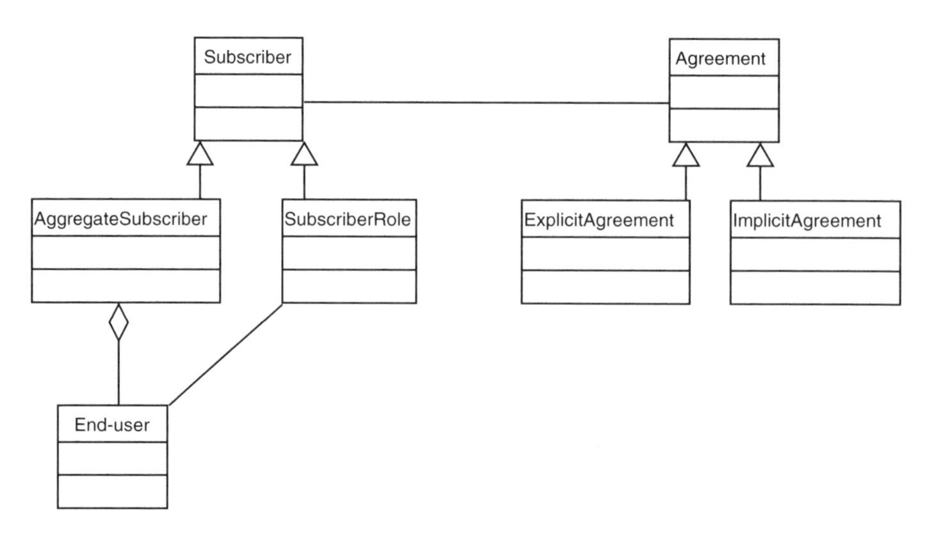

Figure 7.10 Additional customer-related models

back to the service framework, we recall that providers purchase products from one another, and hence are one another's customers.

We shall describe the modelling considerations relating to peer-to-peer services later on in this chapter.

Policy

In this section, policy means a rule specifying that a certain action is carried out when particular criteria are met with. It is assumed that policies can exist at different scopes, so that policies having a wide area of applicability can be complemented by more specific policies. Specific policies can override generic policies, provided that this is allowed in the generic policy. This capability allows the specification of default-type policies for cases in which more detailed policies do not yet exist.

Based on the above, the generic structure of a policy in our context can be described as follows:

```
criteria action priority
```

where criteria tell when the rule is invoked, action tells what is to be done upon invocation, and the priority level can be used for determining which lower-level policies can override a more generic level policy. This system naturally requires the priority management scheme to function correctly.

The concept of overriding rules makes it possible to specify generic rules having a wide area of applicability, yet avoid excessive complexity on the generic level. A simple example of a generic policy could be

```
if (service == VoIP) then TC = conversational, priority=1
```

This rule says that Voice over IP (VoIP) should be allocated conversational traffic class service quality whenever possible. A lower-level policy for a simple Asynchronous Digital Subscriber Line (ADSL) access domain could be

```
if (service == VoIP) then TC = best effort, priority=2
```

reflecting the fact that the domain in question only provides best effort service. Since the latter rule has a higher priority, it overrides the default policy associated with a lower priority.

Policy can relate to legislative aspects, be provider-specific, or be used to express the preferences of end-users as shown in Figure 7.11.

Description of the entities in Figure 7.11:

- Policy: parent class of different kinds of policies
- LegislativePolicy: Policy relating to legislation
- ProviderPolicy: a provider-scope Policy
- EndUserPreference: a Policy for the end-user's needs, covering preferences relating to applications, for example
- ProductPolicy: ProviderPolicy relating to products
- ServicePolicy: ProviderPolicy relating to services
- ResourcePolicy: ProviderPolicy relating to resources
- PrivacyPreferences: EndUserPreferences relating to privacy
- ServiceQualityPreferences: EndUser Preferences relating to service quality.

Legislative policies set the framework within which the service provider must operate.

The provider can have policies which apply to the entire provider domain, as well as product, service, and resource-specific ones. The service policies can be further refined in more detailed policies, as we shall see below.

Two examples of personal policies are given, relating to service quality and privacy preferences. Obviously, there can be many more preferences than just two.

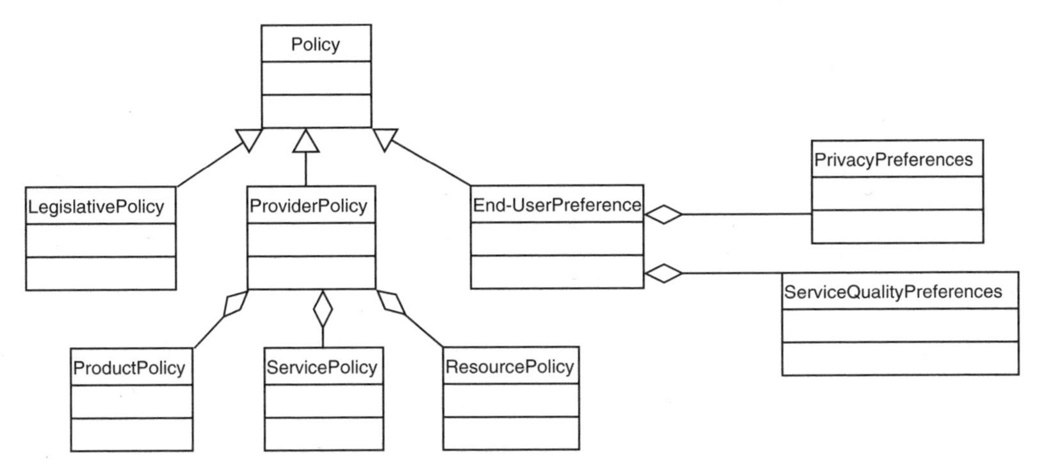

Figure 7.11 Types of policies

Product

We have assumed above that a product is a packaging of services by a stakeholder in a form that can be purchased by another stakeholder.

A product is a part of product offering, and can consist of multiple variants. We shall not address details of marketing-oriented aspects of products, such as the linkage of product offering to business entities. A product can be of a mass-market type or a tailored product. The difference between the two types is in the number of customers using the product. Mobile domain services are often mass-market services, available to millions of users. On the other hand, the providers may have tailored products for large customers. The tailoring of a product can be made easier by having multiple variants of the basic product existing, for example, for different markets. A product is associated with one or more business roles as well as a parameter set.

Basic product-related associations are shown in Figure 7.12.

Description of the entities in Figure 7.12:

- Product: described previously
- ProductOffering: a collection of Products
- ProductConfiguration: parameters relating to a Product
- BusinessRole: role responsible for a Product
- ProductVariant: different variants of a Product
- MassMarketProduct: a Product for mass-market
- TailoredProduct: a Product tailored for a customer
- Subscriber: described previously.

Since the emphasis of this book is on service management, the relatively simple set of concepts described above is sufficient for our purposes.

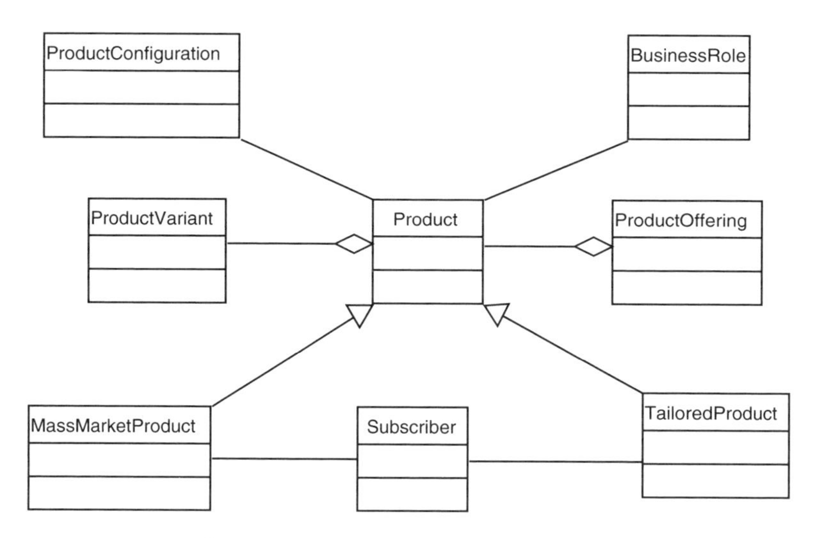

Figure 7.12 Product-related associations

Service

As we have seen previously in this book, the concept of a service is somewhat elusive at the most general level, given the multitude of different viewpoints to all possible services. Fortunately, we need to discuss services in a specific environment, namely, the management of services supporting saleable products. For us, service basically means something that can be instantiated and is associated with a set of parameters.

Products require services to operate. Some services are visible to end-users, whereas some services are internal to the network and not directly visible to customers. The former are often spoken of as end-user services, and are called *Customer-Facing Services (CFS)* in the TMF SID model, in contrast to the Resource-Facing Services (RFS) representing the latter type. The point in separating the types from each other is that the requirements and characteristics of the two types are different from each other. In addition to the shared properties that are common to all services, customer-facing services need to deal with the interfacing towards users and subscribers. The resource-facing services, on the other hand, need to serve as an interface towards underlying resources.

By the above definition, all the services that are not customer-facing are resource-facing in the TMF SID model. This is a clear-cut modelling choice, and as such, easy to grasp. The division also has good justifications for simple systems such as the ones used for providing (VPN) virtual network based services for customers. It has certain consequences, though. It means that the services which are not visible to end-users are always considered to be related to resources. In view of the increasing complexity of services, the division seems somewhat too rigid. The principles of modelling should evolve with the needs of the environment. In the opinion of the author of this book, the only decisive factor should not be whether a service is customer-facing or not. Instead, the basic structure of a model should be sufficient for supporting flexible service life cycle operations.

In this book, we assume that there are three kinds of services. One type represents the capabilities of resources such as the Domain Name System (DNS). We call these resource-facing services. Another type can be interfaced to products, and is called the *product-facing* service. The third type is a category of service that can make use of other services but cannot directly interface to resources or products. We call this abstract service. The addition of abstract services as a third type allows for freedom in the linking of products and services with each other.

Abstract service has been included for multiple reasons. Firstly, keeping resource-oriented services directly linked to resources is clearer and provides a better basis for representing service-oriented architectures with the model. Secondly, having this concept makes it possible to more easily bridge the gap between product-facing services and resource-facing services. Each of the service types can be parameterised. Product-facing services do not need to be always present in peer-to-peer services, and in such cases abstract services can be directly exposed to other parties, provided they are used in an automated fashion without the involvement of end-users and that there are no remunerations involved. It is probably more common for resource-facing services to be operated within a wrapper of some kind, making them abstract services.

Product-facing service can aggregate product-facing services, abstract services and resource-facing services. Abstract service can aggregate abstract services and resource-facing services. Service entity types are illustrated in Figure 7.13. We shall not go deeper

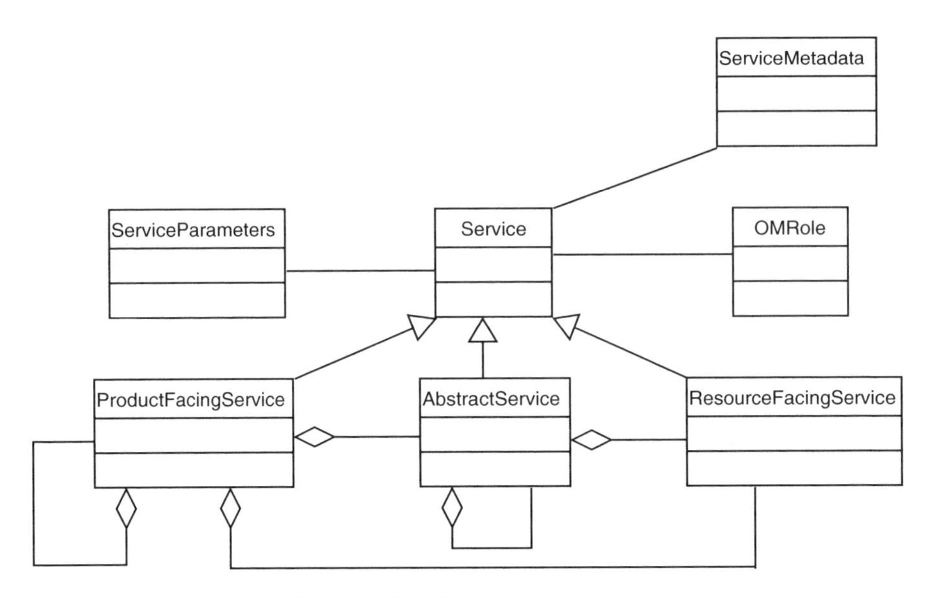

Figure 7.13 Service entity types

into the types of roles involved here, but refer to Figure 5.5 for examples as well as (Service Framework, 2004) and (Koivukoski and Räisänen, 2005) for more information. The figure also shows that services are associated with roles and configurations in the same way as products and resources. Furthermore, service is associated with metadata which tell us how it can be used. As we discussed earlier, this is especially important in SoA-like setting.

Description of the entities in Figure 7.13:

- Service: described previously
- ServiceMetadata: metadata relating to Service, describing how it can be used
- ServiceParameters: parameters of a Service
- OMRole: described previously, in this context responsible for maintaining a Service
- ProductFacingService: a Service that can be linked to products. Can aggregate Product-FacingServices, AbstractServices, and ResourceFacingServices
- AbstractService: a Service which is not ProductFacingService or ResourceFacingSer-vice. Can aggregate AbstractServices and ResourceFacingServices
- ResourceFacingService: a Service which is directly associated with a resource.

Please note that the aggregation of resource facing services is always either an abstract service or a product-facing service.

A few examples of the types of product-facing services are shown in Figure 7.14 in the form of end-user service classes. We shall not attempt to provide an exhaustive list here, but rather to provide a few examples using the division to connectivity services and data transfer services (Räisänen, 2003a). Generally, product-facing services are associated with charging rules in the context of end-user classes they are available to.

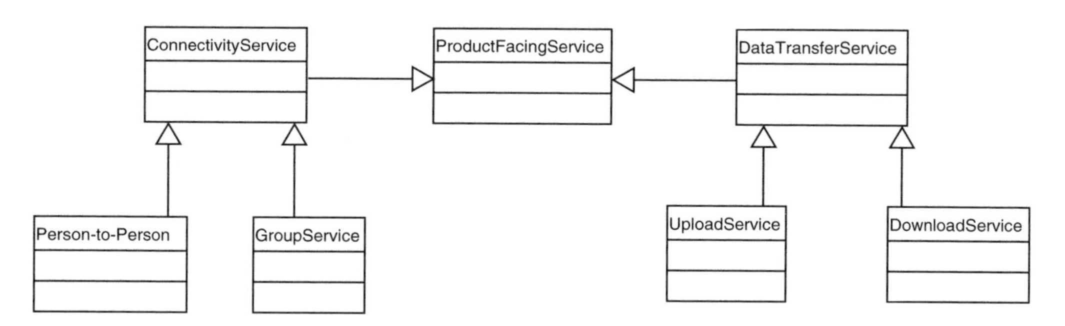

Figure 7.14 Types of end-user services

Description of the entities in Figure 7.14:

- ProductFacingService: described previously
- ConnectivityService: a ProductFacingService relating to connectivity between end-users
- DataTransferService: a ProductFacingService relating to the transfer of data
- Person-to-Person: a ConnectivityService between two persons
- GroupService: a group-oriented ConnectivityService
- UploadService: a DataTransferService for uplink data transfer
- DownloadService: a DataTransferService for downlink data transfer.

The design of resource-facing services follows from the nature of resources involved. Here, it is assumed that resource-facing service conceptually forms an interface to the capabilities resources, roughly along the lines to Common Object Request Broker Architecture (CORBA). The capabilities of the resource must be available via this interface.

Abstract services are constructed for the purpose of making the composition of product-facing services out of resource facing services easier. For example, service management personnel may notice that a particular combination of three resource-facing services is frequently encountered in creating product-facing services. An abstract service may be created to represent the combination of the three services. Another example of the use of an abstract service is creating convenient representations of the capabilities of complex resources or combinations thereof. Multiple abstract services can be created representing a complex resource-facing service, amounting to templates frequently needed in the process of service enabling.

A service may be organised into different variants, reflecting major parameter classifications of the service instances. Service variants can either be used as a conceptual help for understanding the combined effect of service instantiation and service quality support instantiation, or for representing the types of services the difference between which is not merely related to instantiation. Depending on the technological environment, the use of variants can make the control of service instantiation parameters easier.

Some of the fruits of the ongoing work in studying future service platforms in MobiLife and Wireless World Research Forum (WWRF) as well as near-term work in fora such as Open Mobile Alliance (OMA) for example, can already be captured in terms of modelling.

For an added degree of generality, we seek to also cover peer-to-peer services analogously to the MobiLife work cited earlier. One of the important concepts in MobiLife architectural work is support for both managed and non-managed (peer-to-peer) services. On an abstract level, both managed and non-managed services are associated with service providers and service level definitions as shown in Figure 7.15. Service level definition is assumed to consist of service quality level and service level parameters.

Description of the entities in Figure 7.15:

- Service: described previously
- ManagedService: described previously
- Non-managedService: described previously
- ManagedSP: described previously
- Non-managedSP: described previously
- ServiceLevelDefinition: definition of service level associated with a Service
- ServiceQualityLevel: definition of service quality level, a part of ServiceLevelDefinition
- SecurityLevel: definition of security aspects, a part of ServiceLevelDefinition.

Here, we assume that services can either be 'monolithic' or distributed as shown in Figure 7.16. Distributed services require connectivity between the service functionalities the service is composed of and services, whereas 'monolithic' service is assumed to be operated in a single execution environment. It is further assumed that peer-to-peer services can be resource-facing or abstract services, and may also make use of product-facing services. Also, managed services can be distributed. Service functionalities and monolithic services require execution environments to function.

Description of the entities in Figure 7.16:

- Service: described previously
- DistributedService: a Service the implementation of which is distributed in such a way that connectivity between constituents is needed for operation

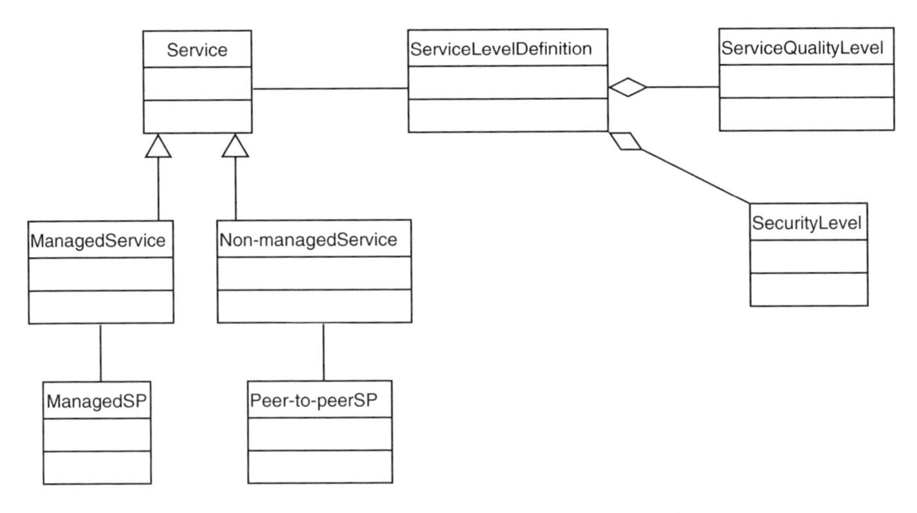

Figure 7.15 Managed and non-managed services

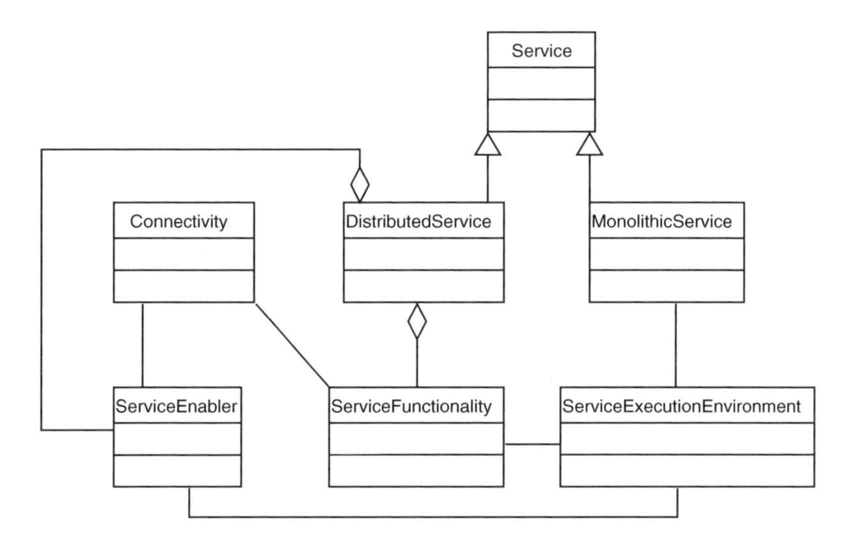

Figure 7.16 Distributed vs. monolithic services

- MonolithicService: a Service which does not require connectivity between constituents to operate
- ServiceFunctionality: a constituent of a Service. Not necessarily a Service in itself
- ServiceExecutionEnvironment: execution environment for ServiceFunctionalities and MonolithicServices
- ServiceEnabler: enabler-type functionalities used by Services. Common to multiple different kinds of Services
- Connectivity: described previously. In this context, supports DistributedServices.

Session

A few meanings for the term 'session' from (OED, 1995):

- A single meeting for (the process of assembly of a deliberative or judicial body to conduct its business)
- A period devoted to an activity.

Both of the above meanings convey the idea of an activity having a well-specified period of time associated with it, and dedicated to a specific task. (The dictionary also mentions 'heavy or sustained drinking' as a specific example of a relevant activity, but we shall not consider this here further). Here we are interested in packet-based services, and define session as being a context for transactions which are created, exist as a unique entity for the duration of the session, and are terminated. A good example of a session in this sense is Session Initiation Protocol (SIP) multimedia session involving multiple participants and supporting dynamic management of service events within the session.

A session is related to an instantiation of certain types of product facing services for a particular usage context. Not all services are session-based. We shall discuss context in a more general manner below, but we have already referred to two aspects of a usage context, namely, end-user and access technology. Generally speaking, a session is related to a product variant and a service variant. Session is associated with a number of service

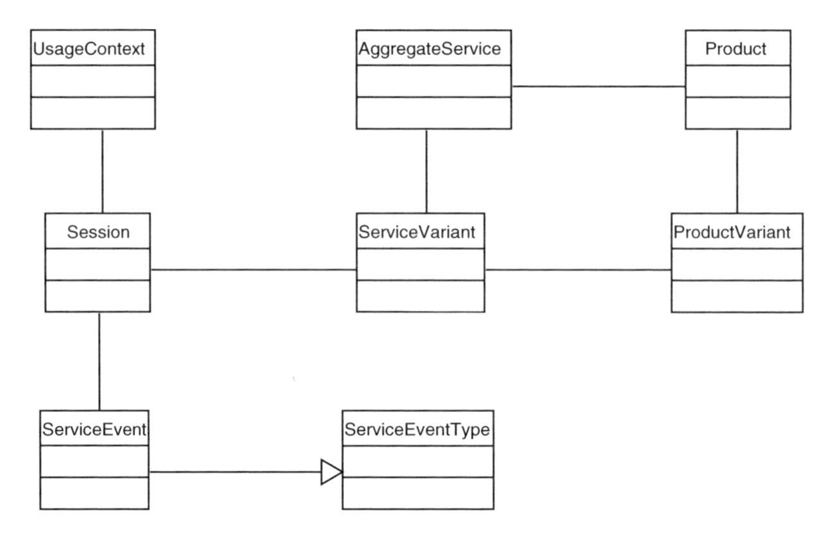

Figure 7.17 Basic session related associations

events, which can be represented as service event types as discussed earlier. Figure 7.17 shows a basic model for these session-related concepts.

Description of the entities in Figure 7.17:

- Product: described previously
- AggregateService: described previously
- ProductVariant: variant of a Product
- ServiceVariant: described previously
- UsageContext: context of usage for a ServiceVariant
- Session: usage session for a ServiceVariant
- ServiceEvent, ServiceEventType: described previously.

Because of usage context, session plays a central role in relation to the service quality and security. To name but a few factors, service quality may be user-dependent so that a certain class of subscriber gets better throughput than others. Available service quality and security features may depend on the access technology and communications endpoints used.

Policies can be used for structuring management information related to service quality and security. Policies can be applied on different levels. In the case of services, aggregate, variant, and service event type specific policies can be identified. A provider may have a domain-wide policy for service quality and security, which may be complemented or overridden by service aggregate type policies. Analogously, service variant level policies can be used to refine aggregate service policies. Service quality and security on a session level are determined based on this policy hierarchy. In addition, service event type policies can be used to complement or replace session-level policies. A simple example of the last is the definition of the default treatment for packets arriving at a network domain. Figure 7.18 illustrates the relation of session to service quality and security using policies.

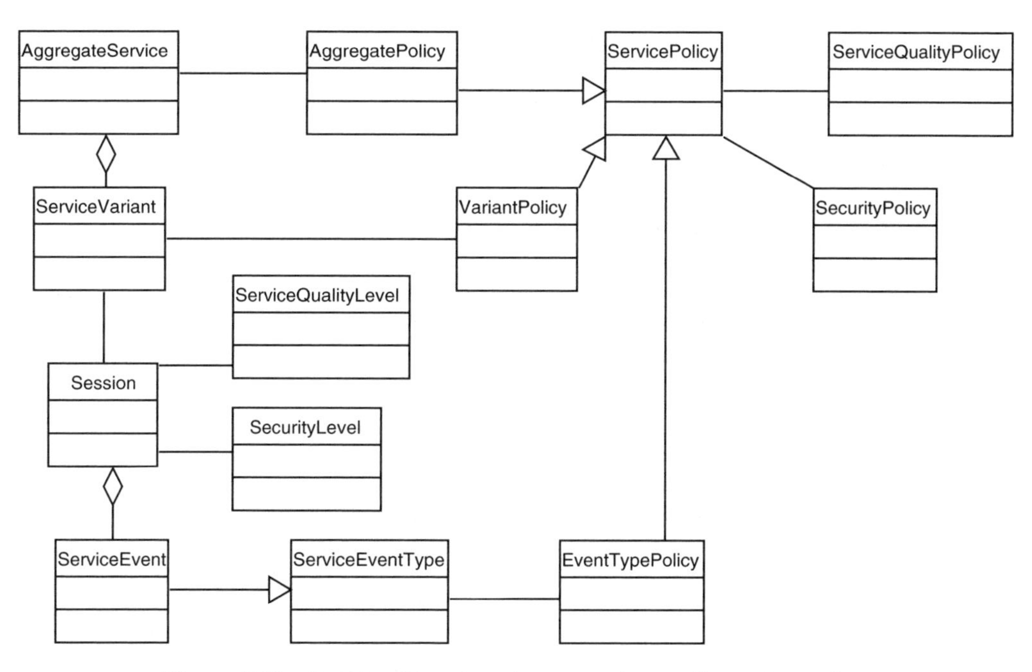

Figure 7.18 Relationship of session to service quality and security

Description of the entities in Figure 7.18:

- ServicePolicy: described previously
- ServiceQualityPolicy: the service quality related aspects of ServicePolicy
- SecurityPolicy: the security related aspects of ServicePolicy
- AggregatePolicy: the policy for an AggregateService, a subtype of ServicePolicy
- VariantPolicy: the policy for ServiceVariant, a subtype of ServicePolicy
- EventTypePolicy: ServicePolicy for a ServiceEventType
- AggregateService: described previously
- ServiceVariant: described previously
- Session: described previously
- ServiceQualityLevel: the service quality level for a session
- SecurityLevel: the security level for a session
- ServiceEvent, ServiceEventType: described previously.

Service event

Service event is part of service variant, and can be used for the structuring of information relating to the technical definition of services. Service quality and security were considered in this context earlier in this part.

Within this book, a service event refers to a specific unit of traffic belonging to an instantiation of a service. Perhaps the simplest example of a service event is an Hyper Text Transfer Protocol (HTTP) reply invoked as a response to a HTTP request. Such an event could belong to remote bank account access, for example. A service event can have also a long duration and consist of a stream of packets, as is the case in content streaming or VoIP telephony media stream. For session-based services, the relation between session and service events is illustrated in Figure 7.17. A single session can include multiple service events of different types.

Fundamentally, a packet-based service event can be modelled as consisting of a number of IP packets as well as a criterion or criteria by which the service event can be uniquely identified. The same criteria are also the basis for service event type classification. As discussed earlier, service event types can be used to operate with classes of service events instead of individual service events. Further, some services require special functionality for creating service events. This functionality is called *service event factory* in our model, and is invoked when service events need to be generated. The resulting model is shown in Figure 7.19.

Description of the entities in Figure 7.19:

- ServiceEvent, ServiceEventType: described previously
- ServiceEventFactory: functionality required to create ServiceEvents
- IPpacket: Internet Protocol (IP) packet, a constituent of ServiceEvent
- ClassificationCriteria: criteria for detecting ServiceEvents or ServiceEventTypes.

There are certain aspects of classification which are of use mentioned in this context, even though we cannot go into detail here. As a background, an IP packet consists of a header section and a payload. The former of these includes information about originating Internet host and destination host, both in the form of unique Internet addresses. In addition, the header includes other protocol-related information, relating to the type of the payload, among other things.

The scarcity of the current version of Internet Protocol (IPv4) addresses has led to the use of Network Address Translation (NAT), which facilitates the multiplexing of multiple IP addresses onto a single IPv4 address which is visible to the outside world. Port numbers are used for differentiating the traffic destined to or originating from IP addresses behind NAT. Because of this, the IPv4 address alone is not always sufficient to identify a communications endpoint uniquely. The use of IP version 6 (IPv6) has the potential to solve this challenge, but IPv6 is not currently widely deployed, except in test networks.

The encryption of Internet content is increasingly used to enhance privacy and confidentiality. VPN is an oft-used form of encryption, where all the traffic, including headers

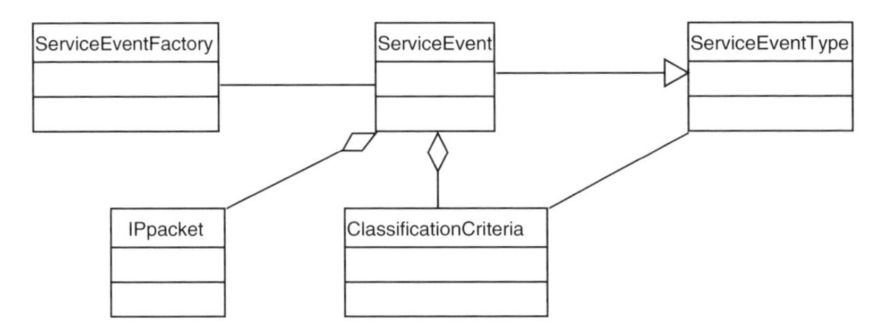

Figure 7.19 Additional service event relations

and payloads, are encrypted between an Internet host and a gateway. Naturally, address information present in the original header cannot be used after encryption. Because of this, a tunnel is set up between the endpoints of a 'tunnel', and an additional header is added to encrypted packets. A consequence of the tunnel set-up, the new headers refer to the endpoints of the tunnel and not to the endpoints of the original communication. In certain situations, this may affect the ability to identify the service events in the original traffic.

Configuration

The configuration of entities is performed as a part of the 'operations', in accord with the service topology defined in the 'design' part of the management framework.

The concept of configuration is used here to represent the parameters needed by an entity to operate. There are configurations related to products, services, and resources. Configurations are related to respective policies, as illustrated in Figure 7.20. No assumptions are made here regarding the mechanism by which configurations are derived from policies. In policy-based systems, the relevant elements may directly use policies. In centrally managed systems, the management system may operate with policies to derive configurations.

Description of the entities in Figure 7.20:

- Configuration: parent class for configurations
- ProductConfiguration: described previously

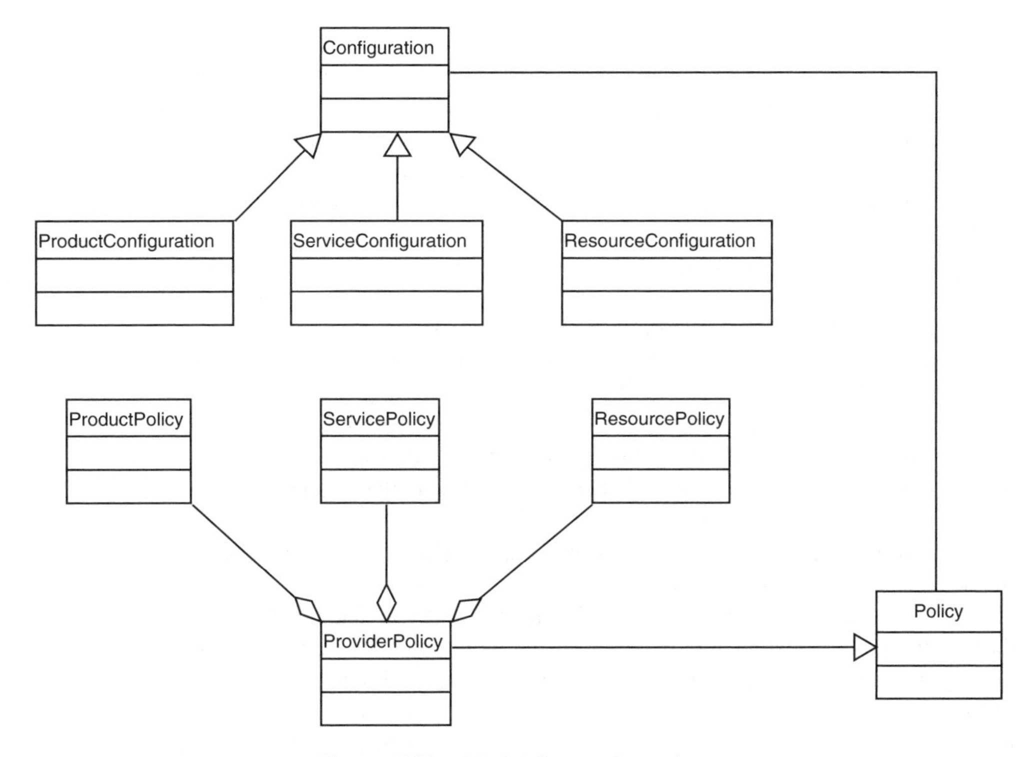

Figure 7.20 Model for configurations

- ServiceConfiguration: described previously
- ResourceConfiguration: a Configuration relating to a resource
- Policy: described previously
- ProviderPolicy: described previously
- ProductPolicy, ServicePolicy, ResourcePolicy: described previously. In this context, Policies based on which Configurations for Products, Services, and Resources are created, respectively.

Logically, configurations form a separate hierarchy from the entities they are related to. Modelling constructs such as aggregation and generalisation can also be used for modelling configurations.

7.2.3 Miscellaneous Patterns

Having covered a number of useful abstract patterns and some additional modelling for the basic entities of service models, we are now prepared to study a heterogeneous set of patterns associated with the typical service modelling tasks. They build on the foundation laid by preceding modelling patterns. Compared to previous patterns, they form a less homogeneous set and are more closely associated with specific tasks. These patterns are considered to be of an exemplary nature.

Charging

Charging and billing are important parts of the construction of modern networks. Due to the use of IP as the basis of communications, it is relatively easy to provide services over the Internet. The challenge lies in accurately charging for the content. Basically, charging and billing can be set up in different ways, assuming that the service provider is separate from the Internet access provider, which can be conceptually assumed to be the case.

One way to model charging and billing is to have a direct relation between the service provider and the subscriber. In this approach, setting up of charging and billing is straightforward from the viewpoint of the service provider. The downside is the subscriber having a number of billing relations with different service providers.

The second option is to have an access network provider to perform charging and billing on behalf of the service provider. This arrangement is frequently used in mobile networks and enhances the usability of mobile services by incorporating all mobile service usage into a single bill. This approach brings with it the need to perform the technical set-up of the arrangement in a manner that is responsive to the business needs of the service provider and the access provider.

Here, we shall describe a number of basic concepts which can be used in both arrangements. Classification-related issues were discussed previously. The basic requirement is the ability to identify the party to be billed for the service. The next question is the amount that is charged from the person in question. The applicable charge may depend on the packaging of the service in the form of products, the packaging, in turn, depending potentially on the class of the end-user. The charging scheme may have variants corresponding to product variants. A simple example relating to the end-user class dependence of product variants would be the inclusion of a set of services into a monthly rate of premier subscription, whereas applying use-based charging for reduced-price subscriptions. The charging scheme includes the tariff and the charging mode. The former specifies rules such as dependencies on the time of day or weekday, and the latter will be discussed next. The basic concepts are illustrated in Figure 7.21.

Description of the entities in Figure 7.21:

- Product: described previously
- ChargingScheme: the charging scheme used for the product
- ProductVariant: described previously
- ChargingSchemeVariant: a variant of ChargingScheme corresponding to a Product
- Tariff: a constituent of ChargingScheme describing applicable tariff
- ChargingMode: the charging mode to be applied for ChargingScheme
- AccessTechnology: the access technology related to a ProductVariant
- End-User Class: the end-user classes relating to a ProductVariant.

Charging mode defines the way communications are charged. The basic possibilities here include event-based charging, session-based charging, time-based charging, and volume-based charging. In event-based charging, charging is based on the number of service events of a certain type, or parts thereof. In session-based charging, the number of sessions is the basis of charging, instead of the number of service events. Time-based charging is quite self-explanatory, and volume-based charging means the charging applied to the number of bytes (or octets) of traffic related to service events. The various kinds of charging modes can be used for different kinds of service events. Volume-based charging is often used in connection with data traffic, whereas time-based charging could be used for viewing music videos, for example. The different types of charging modes can be combined, so that session-based charging can be amended with time-based charging for real-time service events within the session, for example.

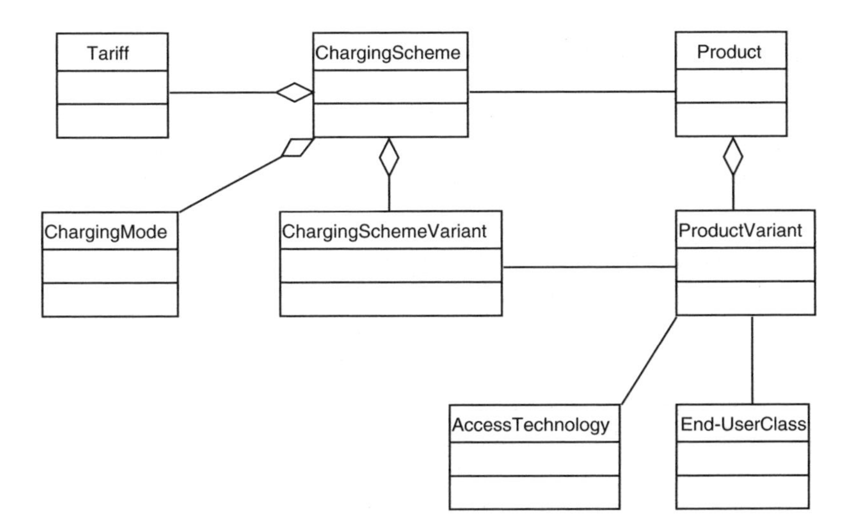

Figure 7.21 Basic concepts for charging

The method used for computing the charge C in each case is not necessarily linearly related to the quantity n which is the basis for charging. In most cases, charging can be modelled on the following form:

$$C(n) = \theta(n, n_0)(An + B) \tag{7.1}$$

where $\theta(n, n_0)$ is zero for $n < n_0$ and one for $n \geq n_0$. Assuming that the basis of charging is time in seconds, free viewing up to one hour and 20 cent charge per minute can be represented as

$$C_T(n) = \frac{\theta(n, 3600)n}{300} \tag{7.2}$$

in Euros or dollars. Similarly, flat rate charging can be obtained with volume-based charging by setting $n_0 = A = B = 0$ in charge computation method, for example. The positive value for parameter B allows for the application of an up-front fee for the first use of a service. In that case, equation (7.1) may need to be amended to have a memory across billing periods, unless the intention is to reset the usage statistics on a monthly basis.

Figure 7.22 shows the inter-relations of charging method entities.

Description of the entities in Figure 7.22:

- ChargingMode: the parent class for charging modes
- ComputationMethod: an algorithm by which charging is determined
- Event-basedCM: an event-based ChargingMode
- Session-basedCM: a session-based ChargingMode
- Time-basedCM: a time-based ChargingMode
- Volume-basedCM: a ChargingMode based on data volume.

Context

The study of the use of context is a hot topic in the field of mobile technology. There are a number of reasons for this. One of them is the ability to create more advanced services. The value of being able to discover the names of the 10 nearest restaurants in a foreign city is indisputable for a frequent traveller. The second reason is the ease of use; by making service better aware of the context of the user, less clicks and other

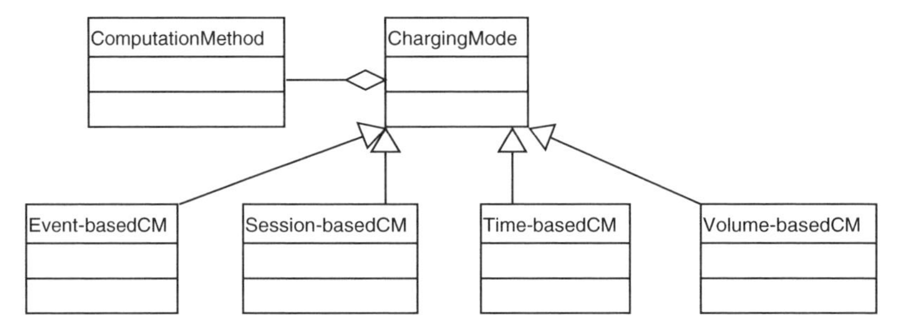

Figure 7.22 Charging methods entities

forms of interaction are needed to achieve useful results. This is especially important in a situation in which an increase in the number of services coincides with the growth of the capabilities and also the complexity of the services.

Location-Based Services (LBS) were already referred to as a part of the restaurant finder example, and are probably the most well-known form of context information. Other forms of context usage, not equally visible to the end-users, include the capability to adjust the content of the messages to the capabilities of the terminal. There can also be multiple variants of the same service for different access technologies, such as General Packet Radio Service (GPRS) and Wireless Local Area Network (WLAN), for example. From a generic viewpoint, the selection of a service variant according to the end-user class could also be viewed as making use of a context, albeit in a rather static form.

From the viewpoint of modelling, context information is used by an instance of a service or service variant. Context information needs to be evaluated in order to be of use for the service. The context management framework is one of the work areas of the MobiLife project referred to earlier. For now, it is sufficient to say that the evaluations can take place in a distributed manner, and participated in by multiple specialised stakeholders. Figure 7.23 shows basic entities potentially related to context. It shows that context requires evaluation functionality to be usable, and has a placeholder for vicinity information as part of the context.

Description of the entities in Figure 7.23:

- Context: a base class for describing a context
- End-User Class: an end-user class relevant to a Context
- End-Point Type: a type of the communications endpoint relevant to a Context
- Location: the location information relevant to a Context
- Vicinity Information: the information about vicinity, acquired by using sensors, for example, that is relevant to Context
- Cevaluation Functionality: the context assessment functionality needed to analyse a Context.

Context information evaluation requires supporting functionality which is not shown here.

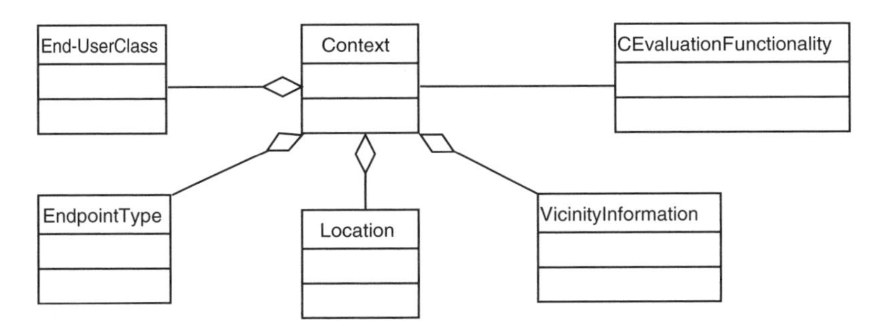

Figure 7.23 Context-related entities

Service level definition

According to the ITU Telecommunication Standardization Sector (ITU-T) Quality of Service framework described in Chapter 6, there are four different perspectives to the quality of service. We shall leave the fourth one aside in the current discussion, and address the first three ones, namely, Quality of Service (QoS) requirements of the customer, QoS planned by the provider, and QoS delivered by the provider. In principle, the end-user experience could also be represented using the framework with suitable extensions required by subjective characteristics. However, our focus is on the technical management of services and we shall leave these aspects out.

The customer has expectations about service quality. Such expectations can be of cross-service nature or service-specific. Here, we shall leave the psychological aspects of QoS expectations aside, and assume that the customer expectations can be expressed formally as service level definition. The degree of detail in customer expectation service level definition can vary, ranging from the definition of a few parameters of interest to detailed specifications of service level definitions.

Providers of managed services are likely to have detailed service level definitions for individual end-user services. A fact discussed in Section 6.5, end-user services are made up of different types of service events, each type having specific requirements and characteristics. Service level definitions are of value in providing satisfactory service usage experience over different kinds of access technologies and usage situations. The definition of the target service level also makes it possible to compare actual service levels to the planned one. Specific views can be created for customers and other providers using role-based access to information, as shown in Figure 7.24. Such views can be used for accessing both service level targets and actual service levels. Analogously to policies, more general entities can be used for providing default values for specific entities.

Description of the entities in Figure 7.24:

- ServiceLevelDefinition: the base class for service level definitions
- CustomerSLD: the customer's ServiceLevelDefinition

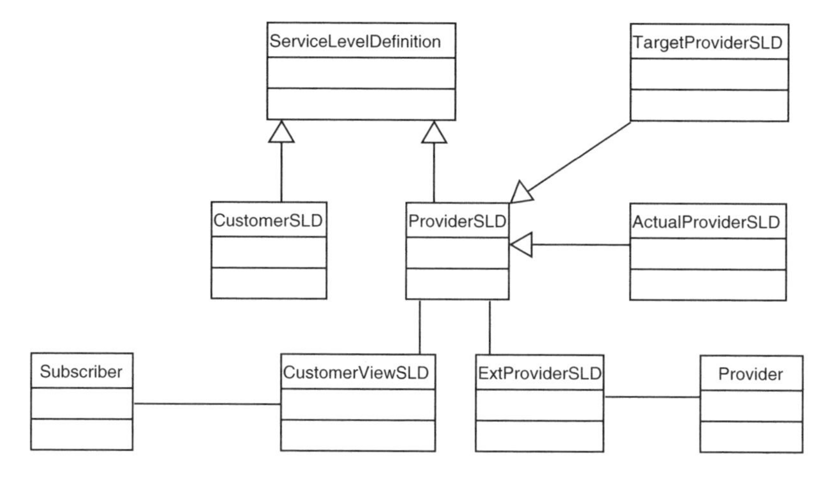

Figure 7.24 Service level details

- ProviderSLD: the provider's ServiceLevelDefinition
- TargetProviderSLD: the provider's targeted service level, a type of ProviderSLD
- ActualProviderSLD: the actual service level, a type of ProviderSLD
- ExtProviderSLD: a view of ProviderSLD for external provider parties
- CustomerViewSLD: the customer view of ProviderSLD
- Provider, Customer: described earlier.

At the most detailed level, service level definition can encompass the different entities of Figure 6.1, as shown in Figure 7.25.

Description of the entities in Figure 7.25:

- AggregateService, ServiceVariant, ServiceEvent, ServiceEventType: described previously
- AggregateSLD, VariantSLD, EventSLD, EventTypeSLD: ServiceLevelDefinitions relating to AggregateServices, ServiceVariants, ServiceEvents, and ServiceEventTypes, respectively.

Non-managed (peer-to-peer) services are – at least conceptually – also associated with service levels, since too poor a service quality also renders peer-to-peer services useless. It is, nevertheless, likely that such definitions are often heuristic or rather loose, unless the platform for peer-to-peer services provides support for automated service level

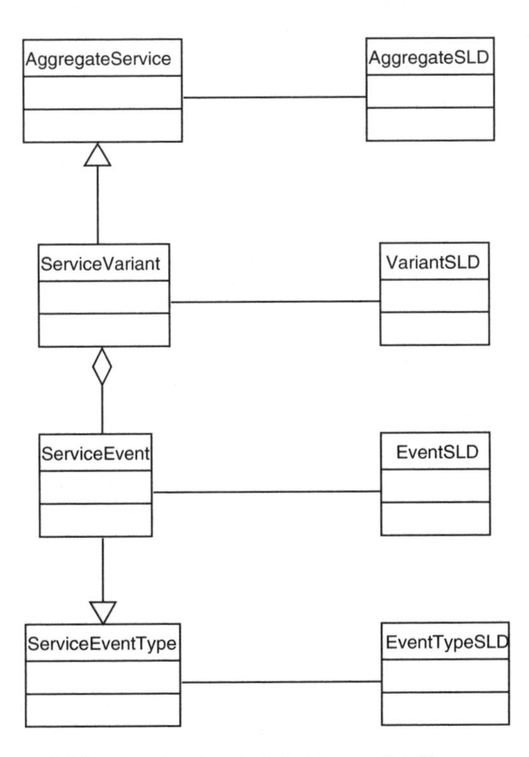

Figure 7.25 Service level definitions of different scopes

management. Addressing this challenge belongs to the working area of research projects such as WWRF and MobiLife.

We shall discuss some of the factors which contribute to end-to-end performance in Section 7.3.2.

Choreography

Choreography is a term which is used in the context of composed services, and refers to the ability to control the composition process. Considered in a broad sense, it needs to support both the static and dynamic composition of services.

It is the task of choreography to make sure that services are put together in the correct manner. Basically, this amounts to using the right ingredients and applying them in the right order.

In the case of dynamic composition, one possibility is to use service metadata directly for the choice of correct constituents. Another possibility is to use service broker systems, should they be available. In the former case, the service registry is needed for discovering the services. In the latter case, choreography can be based on broker-assisted logic. The ontology related discussion in the context of metadata is thus valid for choreography. In more future-oriented scenarios, metadata can include information about dependencies of service functionalities on other ones. Clearly, these scenarios are more challenging in view of the conditional logic needed for choreography.

For static composition, at the simplest level choreography can mean the ability to define in which order service is composed. Metadata can be used for run-time sensibility checks as discussed previously.

Taking into account error conditions is an important issue for choreography. The absence of a specific service functionality can be handled in different ways, depending on the case. For example, an alternative functionality is identified, the functionality is left out of the composition altogether, or composition is cancelled (rolled back). In the first two cases, the characteristics of the composite service may be affected. In the longer run, approximate matching techniques are assumed to be of value.

Assuming that choreography is able to handle conditional logic and metadata, it can be used for the dynamic composition of distributed services. This brings into the picture the service level definition issues discussed earlier. Choreography can take into account service level impacts associated with each service functionality as well as available connectivity instances between them. These data can be made in determining whether a particular collection of service functionalities and connectivities between them meets the end-to-end service level requirements. Different alternatives can also be compared using these data. We shall discuss this further in the connection of distribution.

On a fundamental level, a process can be modelled as being composed of events that exchange messages with each other. An event is associated with triggers, input, actions, and output. The trigger defines when the event is carried out. By the suitable modelling of input, triggers can be defined as consisting solely of inputs. An input, thus, can consist of communications from other processes, or of environmental factors such as time. Action defines what constitutes an event, and can involve other events. Output defines what happens after the action part is finished, and typically involves sending messages to other events. An event can involve conditional logic, so that the action taken can depend on the

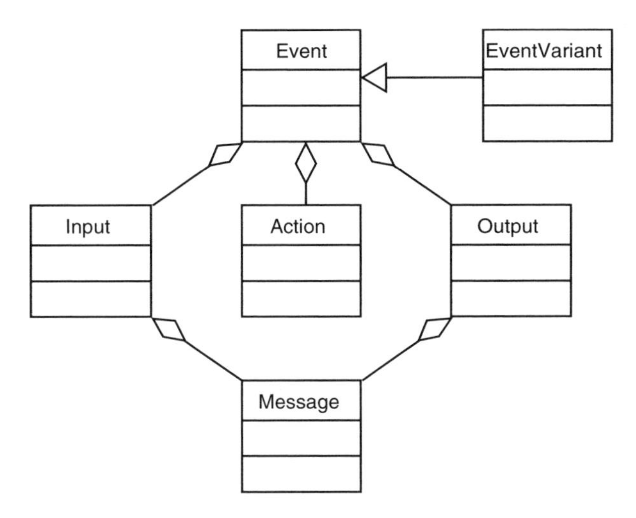

Figure 7.26 Event-based modelling for processes

input, and output can depend on the outcome of the action. The UML model for events is shown in Figure 7.26. In the figure, variants are placeholders for different lines of action carried out in response to the inputs and the results of actions.

Description of the entities in Figure 7.26:

- Event: a base class for describing events
- EventVariant: an event variant, a type of Event
- Input: the inputs relevant to an Event
- Action: the actions carried out during an Event
- Output: the outputs related to an Event
- Message: the messages belonging to Inputs and Outputs.

Note that we have not modelled the relation between Action and Output above. Events can be aggregated using other events, leading to hierarchical designs.

The same kind of modelling can also be applied to business process modelling (cf., e.g. (Hollander *et al.*, 2000)).

Bearer

A bearer is a special case of a connection, and has at least two connection termination points associated with it. A bearer represents the capability to transport information between the termination points with a predefined service level definition. The word 'definition' is to be interpreted here as being of Service Level Objective (SLO) type. For example, a bearer is used in 3GPP and ITU-T analyses as an abstraction layer hiding details relating to underlying technology which are irrelevant for a particular purpose, such as service quality support architectures. Bearers can be aggregated in a hierarchical manner, which is the case in 3GPP definitions for example, where an end-to-end bearer consists of a mobile network bearer, an external network bearer, and a communications endpoint internal bearer. The mobile network bearer, in turn, consists of a radio access bearer and a core network bearer, each of them consisting of lower-layer bearers.

Generally, two kinds of bearers can be identified: connection-oriented bearers and connectionless bearers. Connection-oriented bearers are based on reserving capacity in the intermediate nodes between communication endpoints, and subsequently are typically associated with the stable routing of the connection. Public Switched Telephone System (PSTN) networks are typical examples of this type. For connectionless bearers, only the connection termination points are stable, and – in principle – each packet of traffic can be routed separately.

Due to the different fundamental characteristics of connection-oriented and connectionless bearers, the essence of bearer definition must be devised with a clear goal in mind. The essential commonality between the two types is that they are associated with two or more connection termination points and service level definition. The latter refers to the performance between the communication termination points. Thus, for both types of bearers, the service level can, in principle, be monitored with the measurements performed between the termination points. Please note that this definition does not prevent more fine-grained measurement methods from being used for connection-oriented bearers. A more detailed discussion about bearers can be found in (SoIP business requirements, 2005).

Basic bearer-related entities are shown in Figure 7.27.

Description of the entities in Figure 7.27:

- Connection: described previously
- ConnectionTerminationPoint: the termination points for a Connection
- Bearer: a type of Connection associated with ServiceLevelDefinition
- ServiceLevelDefinition: described previously
- COBearer: a connection-oriented Bearer
- ClessBearer: a connectionless Bearer.

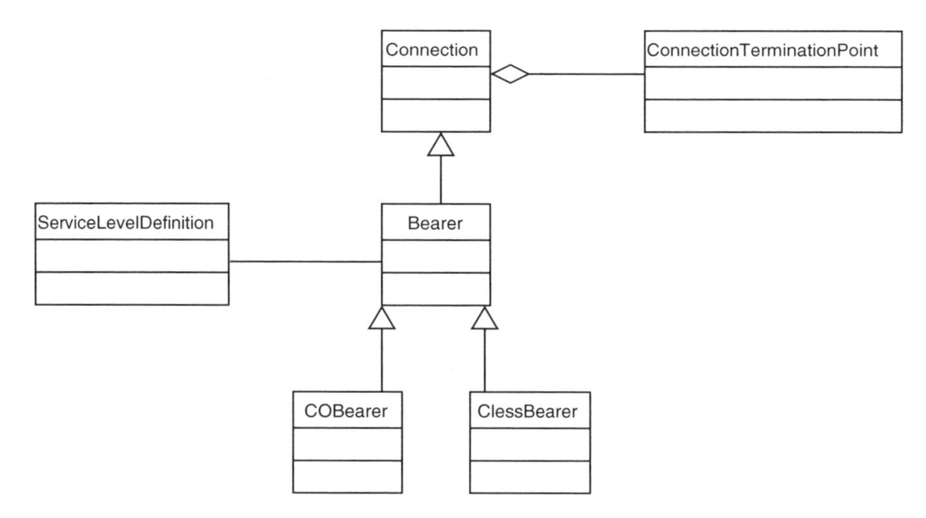

Figure 7.27 Basic model for bearer

Distribution

Distribution in this context refers to the ability to compose service out of service functionalities, which may be physically distributed. We shall not care for the time being whether service composition is static or dynamic in nature. We shall discuss aspects of service composition criteria at a latter stage.

An important factor for distribution is the dependence of the composed service on connectivity and factors relating to individual service functionalities. For purely managed services, the connectivity between service functionalities is most likely managed as well, whereas for peer-to-peer services it can be of an ad hoc, non-managed type. A mixed peer-to-peer/managed service can have both kinds of connections. The functioning of service functionalities – which can be product-facing services, abstract services, or resource-facing services – depends on the service execution environment. The basic relations were shown in Figure 7.16.

The service level target definition has an impact on the distribution of services. Each service functionality is associated with service level impact, which depends on the service in question. Service level impact, in turn, depends on characteristics of the service execution environment, such as processing capability and amount of memory available. Similarly, connectivity between the service functionalities is also associated with service impact. Related entities are shown in Figure 7.28.

Description of the entities in Figure 7.28:

- DistributedService, ServiceLevelDefinition: described previously
- Connectivity, ServiceFunctionality: described previously
- ConnectivitySLI: service level impact related to a Connectivity
- ServiceFuncSLI: service level impact related to a ServiceFunctionality.

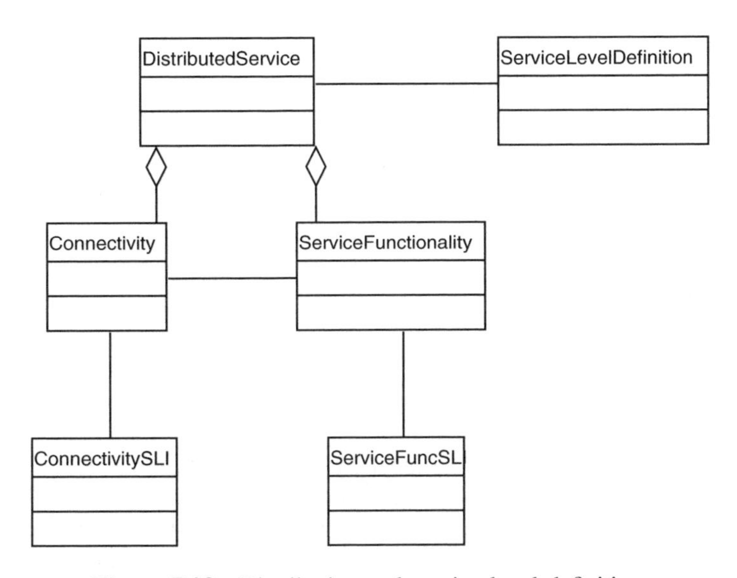

Figure 7.28 Distribution and service level definition

An important aspect of service level impact is its ultimate dynamic characteristic nature. In the case of service functionality, which requires lots of memory or processing power, the load from other service functionality instances affects the service impact. Thus, service impact information should either be dynamically available, or the instantiation of service functionalities on a particular platform should be controlled in such a way that service impact is bounded.

Service impact due to connectivity is another important consideration. Typically, the connection performance information is not available between arbitrary points. In the case of typical backbone, well-formed performance data is available for a relatively small number of peering points. Service impact evaluation logic can be built around estimations based on the closest peering points for managed connections. The service impact assessment in peer-to-peer, non-managed networks is, in the general case, difficult, for the basic reason that no one can provide guarantees or even information about service quality. In the worst case, the topology of the ad hoc network may change at any time.

Peer-to-peer services

The fact that services can be provided by communications endpoints directly in peer-to-peer fashion leads to novel modelling patterns. A participant in peer-to-peer communications – along with necessary equipment, naturally – can be an end-user, service provider, or connectivity provider. The 'or' in the previous sentence is not exclusive or (XOR) of logic, but a peer-to-peer participant can act in different roles simultaneously. In modelling, the concept of role turns out to be useful again here. Each of the roles can be associated with a policy being part of the end-user preferences for peer-to-peer communications. The basic relations reflecting the above discussion are shown in Figure 7.29

Description of the entities in Figure 7.29:

- p2pParticipant: a participant to peer-to-peer communications
- End-User Preference: described previously
- End-User Role: a view on p2pParticipant from end-user perspective
- ServiceProvider Role: a view on p2pParticipant from service provider perspective
- Connectivity Provider Role: a view on p2pParticipant from connectivity provider perspective
- End-User Policy: An End-User Preference relating to the end-user role in peer-to-peer communications
- ServiceProviderPolicy: An End-User Preference relating to the service provider role in peer-to-peer communications
- ConnectivityProviderPolicy: An End-User Preference relating to the connectivity provider role in peer-to-peer communications.

An issue specific to peer-to-peer connections, the forming of a connection typically requires the consent of the endpoints participating in it. The consent can be explicit or implicit in nature. Using the classification of Figure 4.6, peer-to-peer connectivity can be considered to be a type of non-managed connectivity requiring acceptance from participating endpoints. Connectivity provider policies can be used for automating the management

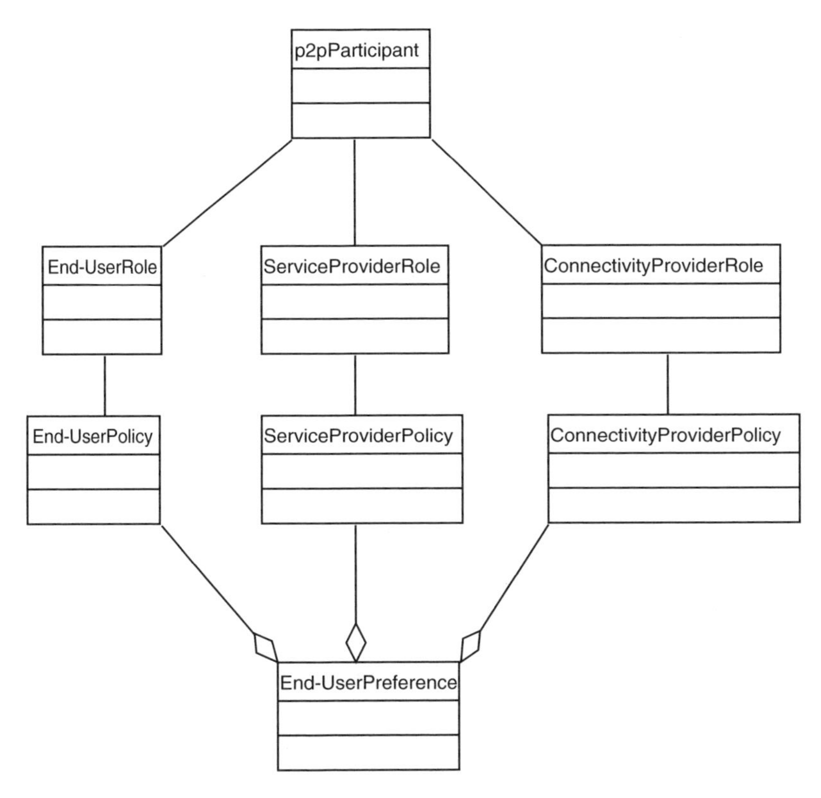

Figure 7.29 Peer-to-peer relations

of consent to communication. Resource-wise, a peer-to-peer connection requires participating endpoints. In case of direct peer-to-peer connection, no other resources are needed.

Conceptually, a peer-to-peer connection has a service level definition associated with it. A peer-to-peer connection is viewed to be a subspecies of non-managed connectivity, the service level definition defines the maximum service level for the connection. Furthermore, the service level definition is dependent on the endpoint capabilities and on connectivity related policies of participants. The relations discussed above relating to peer-to-peer communications are shown in Figure 7.30.

Description of the entities in Figure 7.30:

- Non-managedConnectivity: described previously
- ConnectivitySLI: described previously
- P2PConnectivity: described previously
- P2PSLI: peer-to-peer service level impact, a type of ConnectivitySLI
- Endpoint: representation of peer-to-peer communications endpoint
- ConnectivityConsent: consent to peer-to-peer communications, related to an Endpoint
- ConnectivityProviderPolicy: described previously, in this context can be used for controlling ConnectivityConsent.

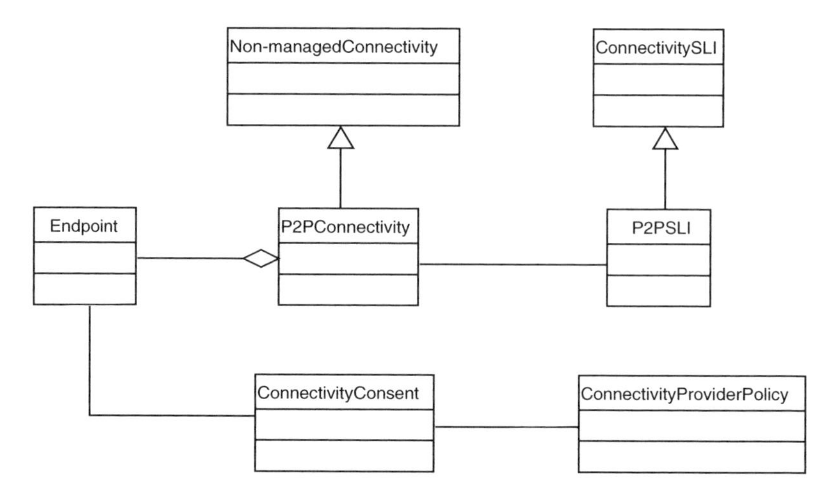

Figure 7.30 Peer-to-peer communications relations

Peer-to-peer services belong to non-managed services, and can also be distributed. Distribution may refer to the distribution within the ad hoc group, or also the inclusion of managed service functionalities or service enablers via wide area connectivity. Peer-to-peer services make use of peer-to-peer connectivity. The two uses of peer-to-peer connectivity can be modelled using service usage role and service composition role, as shown in Figure 7.31. Analogously to connectivity, P2P services can be viewed to have

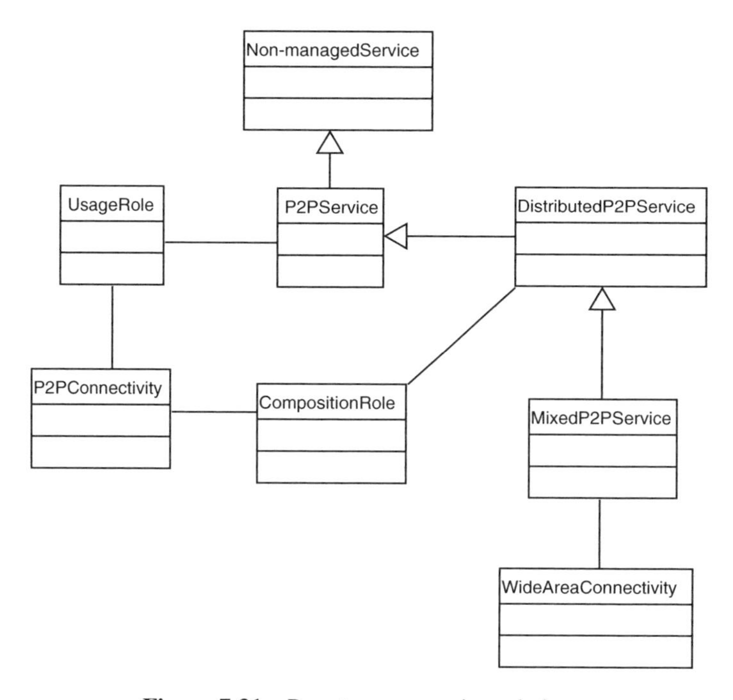

Figure 7.31 Peer-to-peer service relations

a service level definition associated with them. For peer-to-peer services, security related definitions are most likely more relevant than service level related ones.

Description of the entities in Figure 7.31:

- Non-managedService: described previously
- P2PService: peer-to-peer service, a type of Non-managedService
- DistributedP2PService: a P2PService which requires connectivity to service functionalities
- UsageRole: a view on P2PService
- P2PConnectivity: described previously
- CompositionRole: a view on DistributedP2PService related to service composition
- MixedP2PService: a DistributedP2PService which requires service functionalities not present in a peer-to-peer context
- WideAreaConnectivity: a type of Connectivity for accessing managed ServiceFunctionalities or ServiceEnablers.

Privacy

As discussed earlier, the importance of privacy is increasing as applications and technologies get more elaborate. Privacy is closely linked to security. As many other security-related issues, utilisation of privacy in the best possible way can be sensitive to the effects of small details. Examples can be found in (Schneier, 1996). Because of this, the use of policies in general and personal preferences in particular is expected to be a vital ingredient of the use of policies.

Privacy preferences can be divided into two classes, namely, general privacy preferences and role-related privacy policies. General preferences provide the defaults which can be overridden with role-based privacy policies. For roles, defaults can be defined for classes of roles such as human actors and organisational actors, as well as classes of specific actor types such as 'Internet shopping'. These policies can be further refined or overridden for specific actors. Privacy relations are show in Figure 7.32.

Description of the entities in Figure 7.32:

- PrivacyPreferences: described previously
- GeneralPrivacy: general privacy preferences, part of PrivacyPreferences
- RoleBasedPrivacy: role-related privacy preferences, part of PrivacyPreferences
- HumanPrivacy: human communications related to RoleBasedPrivacy
- OrganisationalPrivacy: RoleBasedPrivacy for communications towards organisational entities
- PersonPrivacy: HumanPrivacy related to a specific person
- ActorPrivacy: OrganisationalPrivacy related to a specific organisation.

End-user preferences

In connection with policy, we modelled end-user preferences as a subtype of policy (Figure 7.11). They include preferences relating to service quality and privacy. End-user preferences provide a means of automating service usage, provided they can be used in using services.

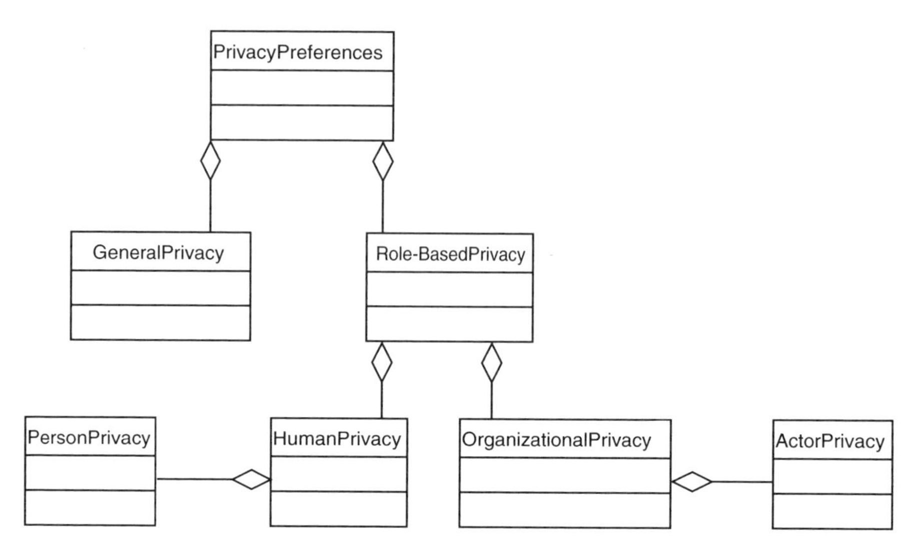

Figure 7.32 Privacy relations

The management of preferences requires the means of accessing them. One or more stakeholders of the service provisioning can provide defaults for preferences which the end-user can utilise as a template in personalising the preferences. Service providers access a subset of the preferences via a specified interface. The end-user is in control with regards to the items for which the preferences are used.

Roles can be used for modelling the different viewpoints to end-user preferences. Additional relations relating to end-user preferences are shown in Figure 7.33.

Description of the entities in Figure 7.33:

- End-User Preference: described previously
- Privacy Preferences: described previously

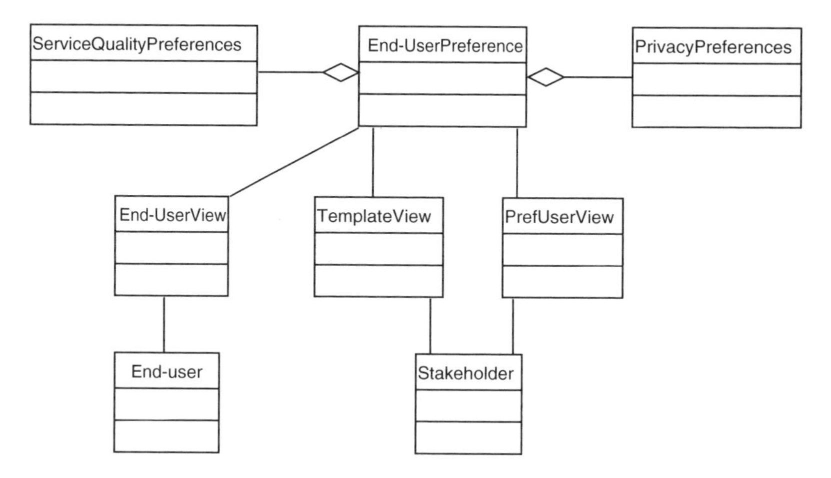

Figure 7.33 Additional relations for end-user preferences

- Service Quality Preferences: described previously
- End-User View: the end-user view on End-User Preferences
- TemplateView: the template for End-User Preferences
- PrefUserView: an external usage view to End-User Preferences
- End-User, Stakeholder: described previously.

7.3 Notes About Using Service Model Patterns

After reviewing a number of service modelling patterns, we shall discuss some issues related to using them in the context of generic real-world problems. This discussion complements the background for service modelling reviewed in Section 7.1. The issues here are partly a summary of the framework, and in part provide novel aspects about its usage.

7.3.1 Using Building Blocks in a Model

The building blocks presented above have been presented 'as is', and there are certain issues to be observed in using them in a service model.

In order to use patterns, a consistent service model needs to be devised. To do this, a process analogous to software design can be used, including actors and use cases. Where service modelling frameworks are used, they constitute boundary conditions for modelling. A service model needs to account for necessary topology (linkage) among products, services, and resources. When reusable components are used, service modelling needs to account for the composition of the components, too.

The inter-relations between entities are rudimentary in the descriptions of the building blocks above, and need to be reviewed in view of application to a specific environment. In addition to relations, one needs to decide which classes can be instantiated directly (are concrete) and which ones cannot (are abstract). The inter-relations between entities must be made more exact than what is included in the task descriptions above; adding names, more detailed descriptions, and multiplicities.

In building a model, one must decide which entities constitute a common model and which ones are domain-specific. Domain-specific models should be viewed to be an instance of the common models. A process for instantiating the domain model also needs to be defined. Aspects of model instantiation have been discussed earlier, including the resolving of dynamic aggregation relations. Another issue that is potentially of importance is the use of inventory for determining the amount of available resources in resilience set-ups on the one hand and for load sharing on the other.

The application of the policy framework described above in a full-fledged manner requires supporting implementation, allowing specification, and the utilisation of policies of various scopes. Policy priority management systems are needed if the multi-priority method described above is to be used.

It is not sufficient to merely develop a model, but the process for maintaining the model also needs to defined. In addition to extending the model, the structure of the model may need to be reconsidered in view of the experiences gained during use.

One of the issues to pay attention to is the relation of service variants, service instances, and service quality experienced by end-users. A particular variant of a service can be instantiated with different parameters. All factors affecting end-user service quality are not always explicitly visible as parameters of the service, but are a consequence of the service quality support instantiation, to use the terminology of (Räisänen, 2003a).

7.3.2 Domain-Specific Requirements

Above, it has been assumed that end-to-end requirements can be specified for a service. Typically, service provisioning also needs to consider allocation of end-to-end requirement characteristics to different network domains. In the context of service quality, it is customary to talk about 'end-to-end budgets' for individual characteristics such as latency, delay variation, and packet loss. The domain budgets of different characteristics require different methods of computing the end-to-end effect; for example, latency is of an additive nature, whereas packet loss requires the application of probability calculus.

Figure 7.34 illustrates the relation between end-to-end and domain-specific requirements. We shall not discuss the details of partitioning end-to-end requirements into domain-specific ones here. An interested reader is referred to (Räisänen, 2003a) for related references. Examples of this type of mapping can be found for delay variation and packet loss (Lakaniemi *et al.*, 2001) and packet loss correlation (Räisänen, 2003b), for example.

Description of the entities in Figure 7.34:

- E2ERequirement: end-to-end requirement
- ServiceQualityLevel: described previously
- SecurityLevel: described previously
- DomainSpecificReq: described previously.

Domain-specific requirements are mapped to relevant service quality support mechanisms. The details of the mapping depend on the technology in question, and are not discussed here. Some examples are provided in Part Three.

As an aside, in the absence of firm service level guarantees, one can also turn the problem around, and view end-to-end service quality as being the product of domain-

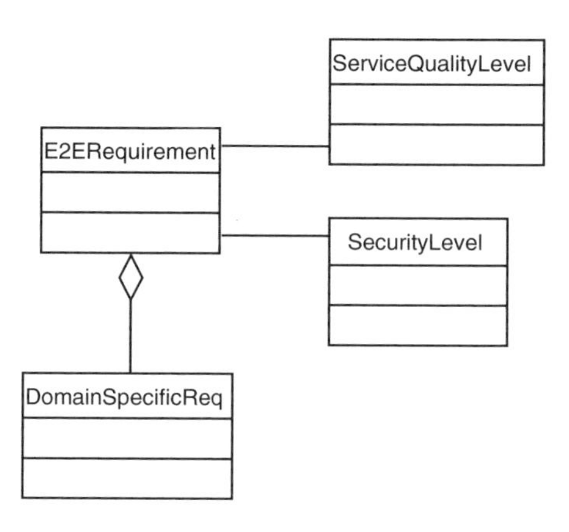

Figure 7.34 Relationship of end-to-end and domain-specific requirements

specific contributions. Using this approach, domain-specific contributions set limits for attainable end-to-end levels.

7.3.3 Service Assurance

Service assurance was one of the primary uses for service modelling identified in the beginning of this book. Measured service quality is already a characteristic which is evaluated by consumer interest groups. There is no question about service modelling being a tool that can be used by service assurance. It should be used effectively to get the maximum benefit out of it, which is our current topic.

Earlier, we described the concepts related to service level definition. Service providers and access providers may have service level definitions relating to target service levels, as well as information about actual service levels. A subset of this information is conveyed to the customers and peer providers as part of SLA related definitions and reporting.

Service modelling is used to link constituent services and resources to the service the performance of which is of interest. In applying the service model for service assurance, simple linkage may not be enough, but special requirements such as the use of adjustable weights in computing contribution to overall service performance may need to be used. It is often necessary to compute service performance in specific locations, which brings geographical context into the picture.

In addition to service performance assessment, service assurance, considered in the broad sense, also includes the management of fault information. It is important to know how much the failure of a specific functionality or a resource contributes to the degree of operation of the overall service. The use of patterns such as resilience and cluster discussed above can be valuable for this purpose.

We shall discuss some related issues in Part Three.

7.3.4 Linkage to Management Framework

Service models need to be linked to the management framework, defining which aspects of the model are available for particular process areas and tasks. The service life cycle of (Service Framework, 2004) provides a skeleton for use in service creation; business, solution, design, implementation and operation views to the model need to be defined. The views define which parts of the model are visible and in which format, and which roles are able to modify and view the model. Similarly, the wider management framework reviewed earlier provides a larger context, allowing of the linking to optimisation.

The service model, viewed in a broad sense of the word, also needs to support linkage to processes. A part of this, the service model needs to define who has the right to modify and access specific data. The model also needs to define who has the right to create, modify, and access the service models. This definition must separately cover the common service models and the domain-specific models. The service model needs to allow for the definition of processes relating to the usage of service models. This typically involves the definition of service management roles and their relations to the information model.

The way of instantiating the model needs to be defined as a set of tasks. Where the model indicates that specific types of resources are needed, a process for defining bindings from the model to the network must be defined. Typically, this process is at least partly automatic, using tools such as network inventory systems, but may also involve humans. Instantiation needs to take into account issues such as logical connectivity between resource.

The linking of service modelling to management may involve multiple stakeholders, and similar roles from different stakeholders.

We shall provide some examples of linkage in Part Three.

7.4 Relationship to Existing Models

After reviewing a set of patterns associated with service management, it is time to step back and review the relation of the approach of this chapter to prior work.

The patterns described above were formulated as a partial answer to the questions listed in the requirements section. As the requirements built partly on the review of related activities in the preceding part of this book, it is clear that the patterns described in this chapter owe a lot to existing work. The TMF SID model is widely known in the area of telecommunications, and has provided plenty of input in devising the patterns here. Object Management Group (OMG), Model-Driven Architecture (MDA) have provided a model for subsequent distributed architectures, and deserve a mention for this reason. Ongoing activities in research projects such as MobiLife and WWRF have provided insight into using distributed service architectures for using context-based services.

In addition to the above 'big picture', a few additional clarifying notes are in order here. The emphasis in this book has been strongly on enabling management of services. This has led to a particular emphasis in modelling. For example, lots of attention has been devoted to service quality level definition. Another example of this, abstract services have been added to complement resource-facing services and product-facing services.

Compared to the TMF SID model, the basic approach is the same, even though we have not aimed at a comparable level of detail and completeness here. This chapter consists of a set of snapshots and guidelines for putting together a service model, instead of presenting an actual model.

7.5 Summary

The patterns described above are building blocks for service models. In building the service models, a clear view of the intended use of the service model must exist. Based on this view, patterns can be reviewed and useful ones incorporated into the model. The use of stakeholders, viewpoints, and other aspects of the Institute for Electrical and Electronic Engineers (IEEE) architecture process can be expected to be valuable in this.

The patterns are snapshots to some of the relevant issues in service modelling. In Part Three, we shall provide examples of using modelling patterns, providing more context and cohesion between the ways that patterns are used.

7.6 Highlights

Ten things to remember from this chapter:

- Service modelling patterns are described.
- Issues such as domain modelling and instantiation need to be considered when putting together actual service models.
- Patterns have been classified into abstract patterns, basic entities, and miscellaneous patterns.
- Role is a powerful pattern, and represents a set of tasks or a viewpoint.
- SLA is a special case of an agreement.
- The end-user can play the role of subscriber.
- Policies can be used by providers and also by end-users.
- Services have been classified into resource-facing services, abstract services, and product-facing services.
- Service can be associated with a service level definition.
- Service level definitions can be associated with aggregate services, service variants, service events, and service event types.

Part III

Use Cases

Scope of Part Three

In this part, we shall illustrate how service modelling concepts developed in Part Two can be used. The application of service modelling patterns and service framework is demonstrated with three examples.

We shall review a DiffServ networking example in Chapter 8 and a mobile network example in Chapter 9. Additionally, we shall provide an example of modelling distributed networking environment in Chapter 10. In each case, we shall provide a description of the example, tell how service framework is applied to the case in question, present a model, and describe how modelling is used by service management processes. The same format of use case description is used in each case, making comparison easier.

The examples are chosen to illustrate application to fixed networking environment, mobile networking, and future-oriented distributed paradigm.

We shall not model stakeholder business models or service provision value nets in the examples.

8

DiffServ Network Example

8.1 Introduction

Our first example is based on the DiffServ networking. The classical DiffServ network was chosen as the technological setting for the first example owing to clarity of the technological set-up. We shall provide below a short description of the functioning of a DiffServ domain, and refer the reader to Appendix B or (Räisänen, 2003a) for longer explanation.

The DiffServ transport network operator provides service quality support within the domain, on the basis of static Service-Level Agreements (SLAs) towards its customers. Traffic conditioning is part of the SLAs. Service quality within the domain is defined using Per-Domain Behaviours (PDBs) pertaining to traffic aggregates. The domain structure of a DiffServ provider can be viewed to be hierarchical, with core routers forming the trunk and spawning branches with edge routers as leaves (Figure 8.1). Finally, let us observe the rather trivial fact that traffic from and to the service provider domain are asymmetrical in the sense that outbound traffic fans out to N egress points, whereas inbound traffic comes in from multiple ingress points.

8.2 Description

We shall use the imaginary end-user service 'Augmented telephony' in our example, adapting and expanding the example from the form used in (Räisänen, 2005). The end-user service facilitates collaboration and consists of the following components:

- Conference management functions
- Voice conferencing bridge
- Group chat
- Document repository.

To add more details to the example, it is assumed that two different end-user classes are defined for the end-user service, namely, a premium price subscription (business) and discounted subscription (nonbusiness). It is assumed here that all the functionalities are available to both end-user classes. Further, it is assumed that usage of group chat and document repository belongs to the basic traffic for business users, but is associated with usage-based charging for the nonbusiness users.

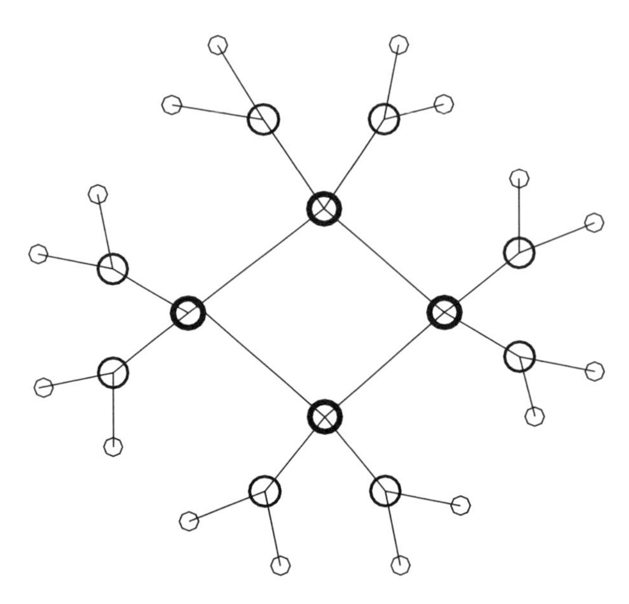

Figure 8.1 Example DiffServ domain topology with edge routers (smaller circles) and multiple levels of core routers (larger circles)

For brevity, it is assumed that all the functionalities are provided by the same organisational party, but this need not be the case always. We shall use the DiffServ framework (Black *et al.*, 1998) of Internet Engineering Task Force (IETF) in this example.

8.2.1 Stakeholders

The stakeholders of the example include end-users, subscribers, connectivity providers, and a service provider. Here, we assume that the service provider is responsible for the entire end-user service, and does not use subcontractors. We shall use only one connectivity provider in the example, but adding more such roles is not difficult. In this example, we assume that the employer of the end-users as subscribers is responsible for all contracts, and the end-users do not play the role of the subscriber. We shall study the consequences of the end-user playing the role of a subscriber in another example later on.

We shall use a relative simple stakeholder configuration for the first example to make it more illustrative and clear. In subsequent examples, we shall study more complex set-ups.

8.2.2 System Description

Below, we shall describe the technical set-ups relating to the service provider and the access provider. Description of the customer will follow.

Inter-stakeholder arrangements

Relations between subscriber and connectivity provider, subscriber and service provider, and connectivity provider and service provider are assumed to be based on traditional SLAs, covering issues specific to DiffServ framework, including Traffic Conditioning

Agreements (TCAs), Service-Level Specifications (SLSs), and Traffic Conditioning Specifications (TCSs). SLAs also define applicable means of monitoring different aspects of the agreement as well as reporting procedures used. SLAs can be explicit or implicit.

SLAs between the service provider and the connectivity provider are assumed to have a clause stating that the terms relating to traffic volumes can be renegotiated periodically.

Service provider

The service provider hosts the execution environment for the augmented telephony service, and has necessary load sharing and resilience arrangements needed for providing carrier-class managed services. The service provider has a direct agreement with the subscriber, and can perform charging within the service domain.

The service provider defines the augmented telephony service on a more detailed level.

Conference management functions include the following:

- Creation of a session
- Adding and removal of participants
- Control of additional functionalities (group chat, document repository).

The voice conferencing bridge works in connection with the conference management, so that users who are allowed to join the conference can also automatically use the phone bridge.

Group chat supports both sending messages to all members of the group, as well as person-to-person messaging.

The document repository has access control that is based on the conference management function, and supports uploading and downloading of content. The documents uploaded to the server are available to everyone in the group, and more fine-grained access control is not considered to be relevant to our example.

It is assumed that the service provider has a Local Area Network (LAN) for system inter-connectivity. For simplicity, it is assumed that the LAN is over-provisioned and does not need service quality support mechanisms for intra-site traffic. Because of the use of traffic conditioning agreements and specifications, the service provider benefits from having the means of supervising and controlling characteristics of outbound traffic aggregates at the LAN egress point towards the connectivity provider. This arrangement supports decisions concerning renegotiation of SLAs by providing information about traffic volumes. Traffic conditioning means such as buffering and dropping can be applied at network egress. We shall discuss some details of traffic aggregation in section 8.3. Depending on TCS, the service provider may not need to perform DiffServ Code Point (DSCP) marking of traffic to indicate how service events are to be treated. Alternatively, the service provider relies on the connectivity provider's capability of classifying traffic.

Connectivity provider

The service provider and customer interface with the connectivity provider's transport networks at points of connectivity. The conformance of inbound traffic to agreed-on profiles can be verified and traffic conditioning and marking can be performed at these ingress points. For incoming traffic from customers, marking is assumed to be determined by the fact that the packets are destined to the service provider's domain and the identification of traffic type based on protocol information in Internet Protocol (IP) packet headers. The points of connectivity, in turn, link to aggregation points such as peering points.

The connectivity provider are assumed to provide service-level definitions in the form of statistical service quality guarantees for traffic aggregates between points of connectivity and selected ingress/egress points. The guarantees make use of the DiffServ PDB concepts such as packet drop rates, throughput, and delay bounds.

The DiffServ framework provides so-called Per-hop Behaviours (PHBs) as traffic aggregates to map flows into. These include Expedited Forwarding (EF) for low-latency delivery and 12 Assured Forwarding (AF) PHBs. The 12 AF PHBs constitute 4 forwarding classes, with 3 drop priorities per class. Forwarding classes can be linked to scheduling, whereas drop priority indicates which packets are disposed of first in case of buffer starvation due to congestion. In addition to the EF and AF groups, Best-Effort (BE) scheduling is usually recognised using the capacity left from other PHBs. The connectivity provider performs DSCP marking at the network ingress, and bases service quality support in the core routers on the PHB indicated by DSCP.

In addition to the implementation of the DiffServ framework, the connectivity provider can use traffic engineering tools such as Multi-Protocol Label Switching (MPLS), for example, for routing control. These are not relevant for our example.

The connectivity provider uses suitable measurements for supervising service quality within its domain and uses these data as a basis for reports specified into SLAs. It is assumed that the measurements are available at PHB granularity between pairs of major points of presence.

Depending on TCAs, the connectivity provider may perform traffic conditioning at domain egress points.

Customer description

The subscriber operates a LAN network for connecting end-users to the connectivity provider. Service quality support mechanisms such as the Institute for Electrical and Electronics Engineers (IEEE) 802.1Q can be used for ensuring adequate service quality for voice in the LAN. We shall not consider customer-internal service quality support here.

The subscriber has SLAs towards the connectivity provider and the service provider as part of business agreements. Towards the connectivity provider, the subscriber also has TCA and related specifications. The customer does not perform DSCP marking of traffic, but relies on the connectivity provider for this instead. Depending on the SLA with the provider, the customer may perform traffic conditioning at egress points.

8.3 Service Framework

We shall now proceed to describe how service framework is employed within the example at hand. We shall do this by discussing how aggregate service, service variants, service events, and service event types are used in our example. We shall concentrate on service quality and security, and handle charging separately.

What is shown below is the service provider's way of using the service framework to define service quality for the service. It describes the planned characteristics of the end-user service, and makes references to DiffServ aggregate mappings in the connectivity provider's domain. It is assumed that adequate aggregates have been chosen on the basis of the PDB definitions supplied by the connectivity provider. The connectivity provider, in turn, uses the mappings defined in the service framework within its domain.

8.3.1 Aggregate Service

The end-user service is defined as consisting of two variants, business service and non-business service. The difference between the two variants is in the repository up/download throughput. The availability target of the overall end-user service is defined as 99%, and the default service quality for service events is PHB Assured Forwarding class AF22. The collection of overall service usage statistics is defined on the aggregate level.

- Business oriented parameters
  - Geographical coverage of the service: entire access network
  - Validity period of the end-user service: until further notice
  - Two variants: business and nonbusiness
- Service quality requirements
  - Availability: 99%
  - Default PHB = AF22
- Service quality characteristics
  - Overall usage statistics of the end-user service shall be stored
- Security
  - All service events are encrypted by default.

As we see above, the aggregate service definition includes the technology-specific parameter (PHB definition).

8.3.2 Service Variants

We shall now describe the business and nonbusiness variants. The business variant has a higher performance than the nonbusiness one.

Business

The composition in terms of service events is as follows:

- Session management signalling
- Telephony media stream
- Group chat
- Data up/download, high throughput.

Parameters include the following:

- Business oriented parameters
  - Access technology: Fixed IP
  - End-user group: business

- Technical parameters
  - Telephony media streams are created using provider's Session Initiation Protocol (SIP) proxy
- Service quality requirements
  - Service instantiation time: 2 seconds
- Service quality characteristics
  - Usage.

Nonbusiness

Composition is the same as for the business variant, but with lower throughput for document repository.

The service quality parameters are the same as for the business variant.

8.3.3 Service Events

Next, we shall discuss the service events that belong to our example end-user service. They include session management signalling, telephony media stream, group chat events, and data up/downloading (two variants).

Session management signalling

This service event category includes traffic that is related to setting up and tearing down of sessions, and adding and removal of participants. Here, we are assuming that adding a session participant also includes setting up of an audio stream for the participant in question. Thus, no separate telephony control signalling service event is required. (Carrier-class IP telephony puts quite strict limits for telephony signalling, but here we are assuming a bit more relaxed approach to make things simpler.)

Regarding service quality parameters, session management signalling should be responsive. The following parameters relate to individual requests and replies:

- Service quality requirements
  - Service quality of designed type
  - End-to-end delay interactive
  - Packet loss low to avoid delay due to retransmissions
- Service quality characteristics
  - Traffic patterns: temporally randomly generated events within a session. Most likely more activity at the beginning of the session than in the middle.

Telephony media stream

Telephony media in our case consists of two periodic transmissions of packets: one or more to the group 'bridge' server from the persons who are talking (hopefully not too many at a time) and a stream from the bridge to each of the participants distributing the common voice signal.

- Service quality requirements
  - The service quality requirements are of inherent type
  - End-to-end delay requirements: less than 400 ms including dejitter buffering at the receiver
  - Delay variation requirements: less than 30 ms as measured at the receiver
  - Packet loss requirements: less than 2% overall, low loss correlation
- Service quality characteristics
  - Token bucket parameters: maximum bit rate capacity reserved
  - Uplink and downlink traffic patterns: both uplink and downlink must be available all the time.

Above, constant availability of the speech signal amounts to effective statistical capacity reservation. Regarding end-to-end requirements, it is useful to note that the endpoint also contributes to the end-to-end values (e.g. dejittering buffer impact on latency).

Group chat

Group chat consists of uplink messages for an individual participant, which are distributed to other participants. Group chat control messages are considered to belong to the same event as the content of the chat messages.

- Service quality requirements
  - Type of requirements: designed
  - End-to-end delay requirements: relatively low
  - Packet loss requirements: relatively low to avoid retransmissions
- Service quality characteristics
  - Uplink and downlink traffic patterns: random.

Data up/download, low throughput

Data uploading and downloading are assumed to consist of individual FTP or Hypertext Transfer Protocol (HTTP) data transfer transactions. Token rate is used to control the average rate, and token bucket size controls the maximum allowable 'bandwidth burst' size. It is assumed that traffic conditioning is performed in the storage server.

- Service quality requirements
  - Type of requirements: designed
  - End-to-end delay requirements: can be relatively long
  - Delay variation requirements: should be low
  - Packet loss requirements: preferably low to preserve throughput
- Service quality characteristics
  - Token bucket parameters: token rate 256 kbps, bucket depth = one MB
  - Uplink and downlink traffic patterns: temporally random one-way transfers.

Data up/download, high throughput

As with data up/download, low throughput except that for token bucket parameters, token rate is 512 kbps.

8.3.4 Service Event Types

Service event types are assumed to be used for mapping packets entering the domain to particular DiffServ PHBs. Three service event types are used: real-time traffic, interactive traffic, and data transfer.

Real-time traffic

- Technical parameters
  - Aggregation criterion: detect RTP traffic
  - Traffic conditioning method: dropping
  - Token rate sufficient for used speech coding scheme
- Service quality requirements
  - Map to EF PHB.

Above, the use of buffering as the traffic conditioning method is dictated by the challenging end-to-end latency requirements. In computing token rate, the effect of RTP and IP headers must be taken into account.

Interactive traffic

The interactive traffic service event type supports the events that do not need as short a latency as real-time events do, but should have better performance than background data transfer.

- Technical parameters
  - Aggregation criterion: detect session management signalling and chat messages
  - Service event type specific service quality policy
  - Traffic conditioning method: dropping or buffering
- Service quality requirements
  - Map to AF11 PHB.

The interactive traffic uses AF11 PHB, which has the lowest dropping priority (i.e. is dropped last in case of congestion). It is assumed that AF class 1 has sufficient bandwidth to accommodate timely forwarding of messages.

Prioritised data

This service event type is for prioritised data that does not require interactive responsivity but should receive better treatment than best-effort traffic.

- Technical parameters
  - Aggregation criterion: detect data up/download
  - Service event type specific service quality policy
  - Traffic conditioning method: dropping or buffering
- Service quality requirements
  - Map to AF21 PHB.

Prioritised traffic uses AF21 PHB, which has a different forwarding treatment from AF11, but has low drop priority.

Bulk data transfer

Background data transfer is used for bulk data transfer, and traffic that has no other rules is mapped onto it.

- Technical parameters
  - Aggregation criterion: any incoming traffic
  - Traffic conditioning method: dropping or buffering
- Service quality requirements
  - Map to BE PHB.

8.3.5 Note

Above, we assume for simplicity that the telephony media stream is mapped onto the EF class without a resource control functionality such as gating. The IP Multimedia Subsystem (IMS)-style authorisation to real-time service quality support can be circumvented, since the voice signal is distributed by the telephony bridge and hence can be detected.

For end-to-end Voice over IP (VoIP) telephony this is not optimal in all cases. For example, the customer LAN may have too much simultaneous voice traffic. Tackling this would require adding resource availability based admission control to the connection set-up.

The default PHB specified on the aggregate service level is overridden for all service event types. The default PHB could be due to provider-level defaults, and as such defined by a separate role than the PHBs of individual service event types.

8.4 Service Model

Below, we shall illustrate aspects of service modelling of our example using views. The model does not seek to be complete, but concentrates mostly on the use of the service framework for our first example.

8.4.1 Use Case View

Use case view for selected service model-related use cases is shown in Figure 8.2. Four use cases are shown: end-user service level definition, PDB definition, service mappings, and actual service level. Note that the first use case is related to the use of the service model by provider stakeholders, and as such represents a 'meta-level view' to the service model. Similar use cases could be drawn for the use of the service, but these are omitted in the interest of brevity here.

Description of the entities in Figure 8.2:

- Service provider: described in Part Two
- End-user: described in Part Two

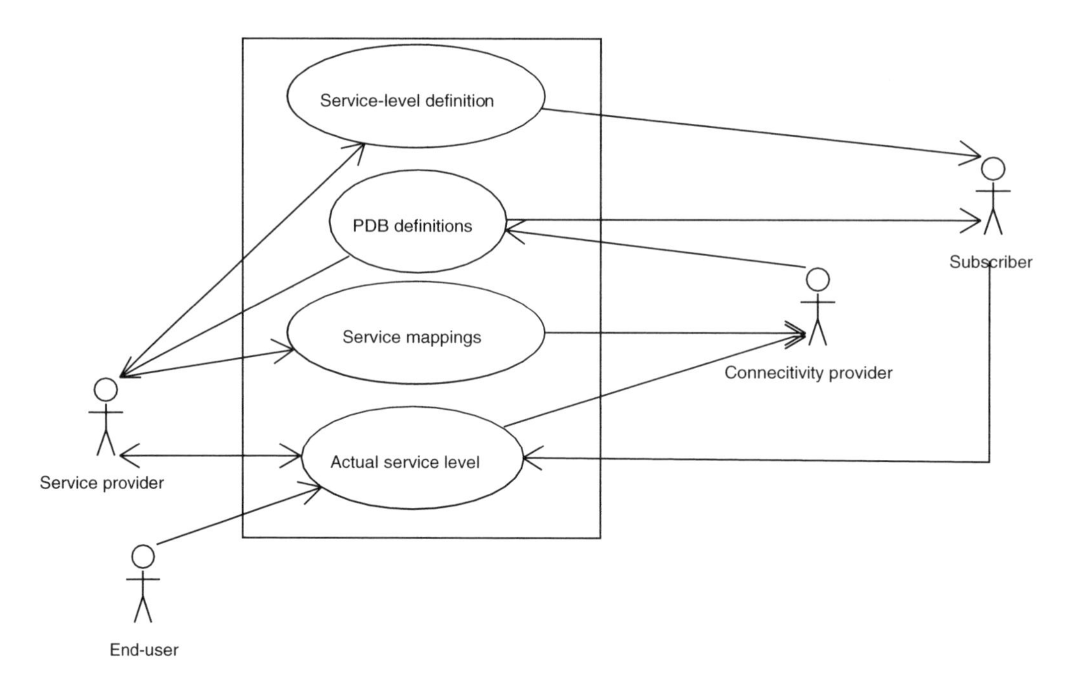

Figure 8.2 Use case overview for DiffServ networking example: service model provider use related use cases

- Subscriber: described in Part Two
- Connectivity provider: described in Part Two
- Service-level definition: the use case for defining target service level for end-user service. Done by the service provider for augmented VoIP and used by the subscriber
- PDB definitions: the use case for defining DiffServ PDBs. Done by the connectivity provider, used by the service provider and the subscriber
- Service mappings: the definition of the mapping of service events to the traffic aggregates. Done by the service provider, used by the connectivity provider
- Actual service level: the use case for information related to actual end-to-end service levels. Information provided by the end-user, the subscriber and the service provider; used by the service provider and the connectivity provider.

Please note that in addition to the provider of the end-user service, the connectivity provider provides service. Thus, information about actual service levels is used by both the provider of the end-user service and the connectivity provider.

8.4.2 Static View

We shall illustrate the use of the static view of service modelling by viewing selected aspects of the system, including composition of end-user service, service quality, and charging. The intention is not to provide a complete model for the example.

We shall begin the description of the static view by presenting a model of the types of services involved. Services include an end-user service (augmented telephony) and

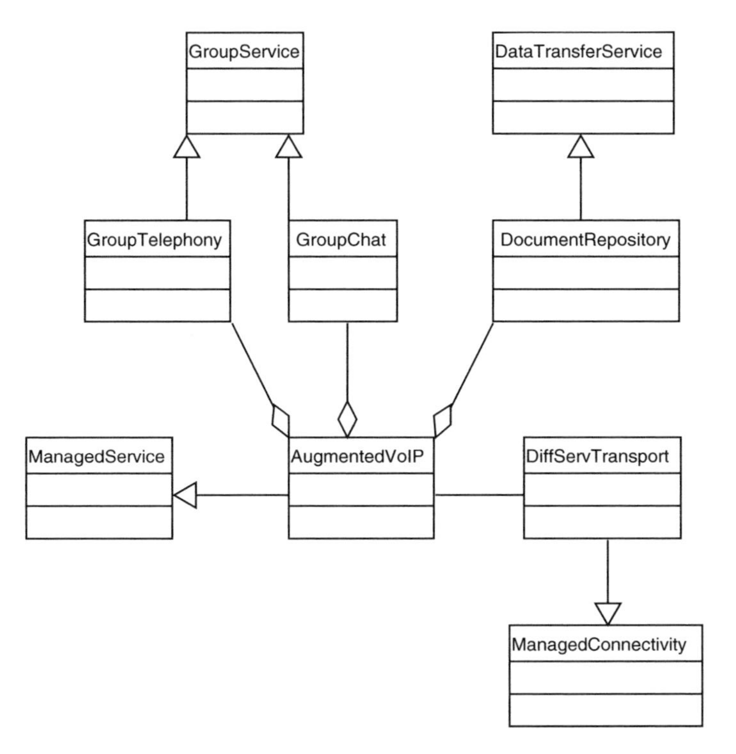

Figure 8.3 Basic relations of augmented telephony service

transport service, which is used by the end-user service. The basic 'genealogy' of services making up the augmented telephony is shown in Figure 8.3. In the model shown, the product 'augmented telephony' aggregates product-facing services.

Description of the entities in Figure 8.3:

- GroupService: described in Part Two
- DataTransferService: described in Part Two
- GroupTelephony: the representation of IP telephony bridge service
- GroupChat: the representation of group chat service
- DocumentRepository: the representation of document repository functionality
- AugmentedVoIP: the representation of augmented VoIP service consisting of group telephony, group chat, and document repository
- ManagedConnectivity: described in Part Two
- DiffServTransport: DiffServ-based ManagedConnectivity. Used by AugmentedVoIP in this example.

DiffServ transport is also the basis for the connectivity service for the end-user service provider and subscribers. From that perspective, the connectivity provider also provides a service. Related modelling is shown in Figure 8.4 using roles for representing the aspects of connectivity service relating to end-user connectivity and service connectivity separately. These can be viewed to constitute business-to-customer and business-to-business

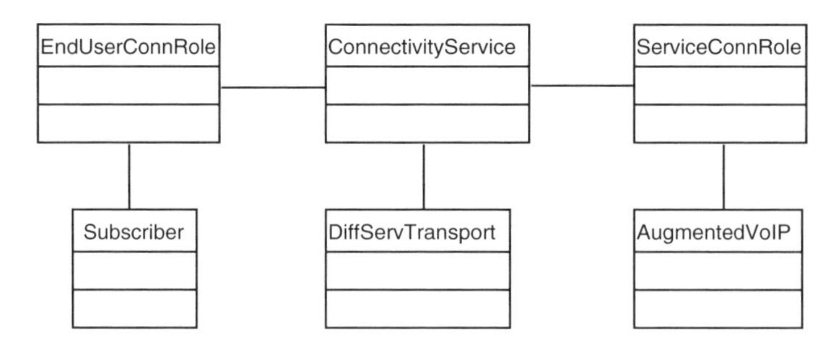

Figure 8.4 Connectivity related modelling for the first example

views, respectively. Note that this also means that the connectivity provider plays the role of a service provider towards both the customer and the end-user service provider.

Description of the entities in Figure 8.4:

- ConnectivityService: described in Part Two
- DiffServTransport: described previously
- Subscriber: described in Part Two
- AugmentedVoIP: described previously
- EndUserConnRole: the end-user connectivity oriented view on ConnectivityService
- ServiceConnRole: the service provider oriented view on ConnectivityService.

The composition of the augmented VoIP in terms of service events is depicted in Figure 8.5. Service events that are common to both variants of the service are represented as being associated with the aggregate service, whereas variant-specific events are linked to variants. Please observe that the modelling shown is analogous to the use of virtual

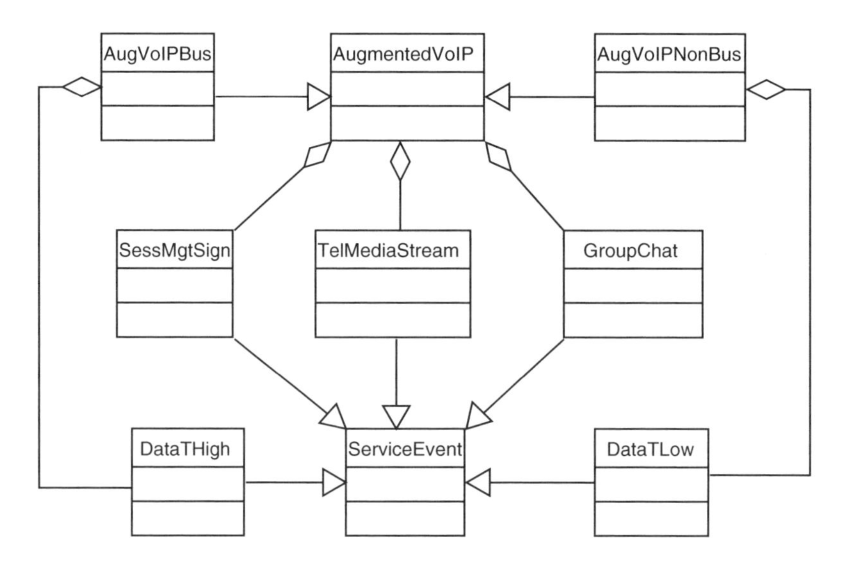

Figure 8.5 Model for the composition of augmented VoIP service

functions in C++ programming language; actual functionality is not at all described for parent class, only for inherited objects. An alternative approach would be to define, for example, DataTLow as the default in AugmentedVoIP, and then override it for AugVoIP-Bus. The modelling shown in Figure 8.5 befits a workflow where the service designer leaves the specification of data transfer service events for other roles.

Description of the entities in Figure 8.5:

- AugmentedVoIP: described previously
- AugVoIPBus: the business variant of AugmentedVoIP service
- AugVoIPNonBus: the nonbusiness variant of AugmentedVoIP service
- SessMgtSign: the session management signalling service event
- TelMediaStream: the telephony media stream service event
- GroupChat: the group chat service event
- DataTHigh: the high-throughput data up/download service event
- DataTLow: the low-throughput data up/download service event.

Figure 8.6 shows mapping between service events and DiffServ traffic aggregates discussed earlier. Note that there is also a placeholder for unknown traffic events, which are mapped to best-effort class. Also please note that we only need a single interactive traffic event type, since we assume that traffic shaping is performed at the service domain for data up/download.

Description of the entities in Figure 8.6:

- TelMediaStream, SessionMgtSign, GroupChat, DataTHigh, DataTLow: described previously
- UnknownEvent: the event that belongs to none of the above classes

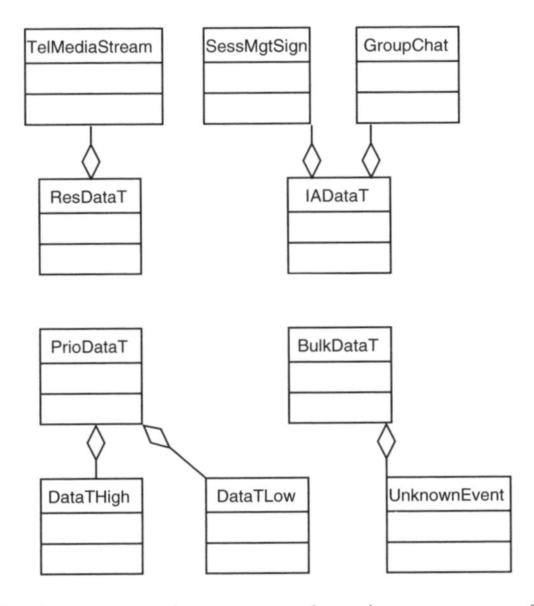

Figure 8.6 Mapping between service events and service event types for augmented VoIP

- ResDataT: effective capacity reservation data transfer service event type
- IADataT: interactive data transfer service event type
- PrioDataT: prioritised data transfer service event type
- BulkDataT: BE data transfer service event type.

The charging scheme depends on the variant, according to the description provided in Section 8.2. To illustrate different charging modes, we assume that there is a per-session charging applied to the service, charged from the arranger of the meeting. Each participant to the telephony session is charged according to time-based charge. For non-business variant users, group chat is charged on the basis of the number of events and data up/downloading according to traffic volume. The model for charging is shown in Figure 8.7. Roles are used for describing different charging aspects of variants on one hand, and for representing the charging applied to the group controller on the other.

Description of the entities in Figure 8.7:

- AugVoIPBus, AugVoIPNonBus: described previously
- SessMgtSign, TelMediaStream, GroupChat, DataTLow, DataTHigh: described previously
- Session-basedCM, Volume-basedCM, Event-basedCM, Time-basedCM: described in Part Two
- GroupControllerRole: the role representing the session participant who controls the role

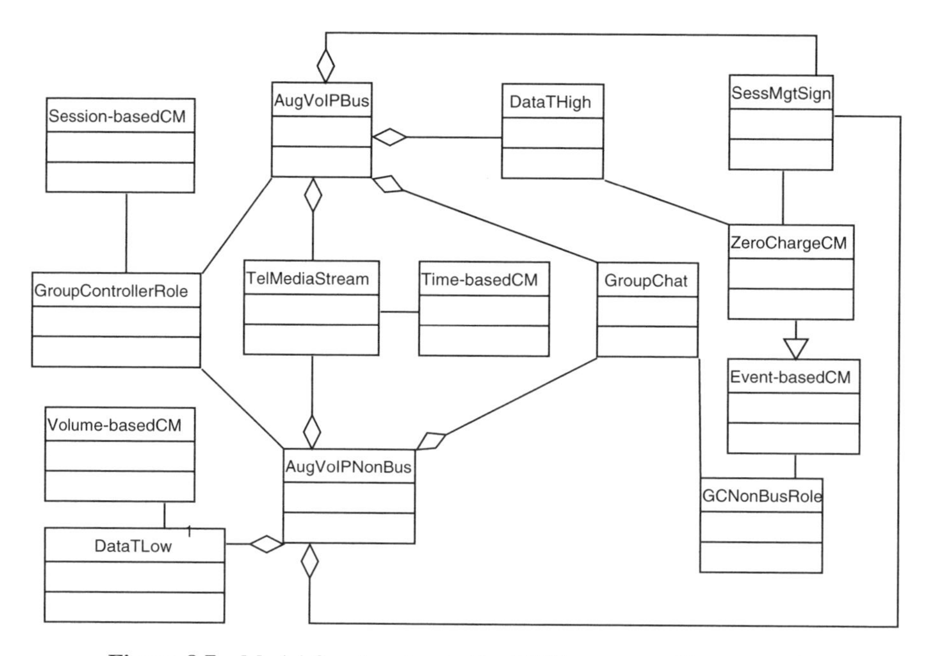

Figure 8.7 Model for charging in the DiffServ networking example

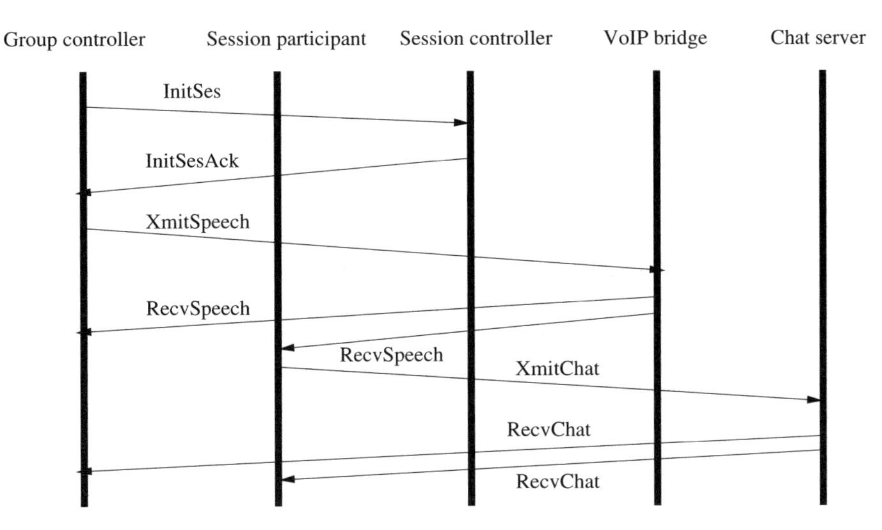

Figure 8.8 A dynamic view for the DiffServ networking example

- GCNonBusRole: the role representing charging aspects of nonbusiness variant of group chat
- ZeroChargeCM: a variant of Event-basedCM representing zero charging.

We shall not consider mapping to resources for this example. An example of this will be provided in the next chapter.

8.4.3 Dynamic View

We show an example signalling sequence relating to a service usage session. It is depicted in Figure 8.8. The signalling sequence shows a part of a session, consisting of session set-up, use of VoIP bridge to deliver speech from the group controller to the other participants, and group chat related events for a chat message sent by a participant.

Description of the entities in Figure 8.8:

- Group controller: the party controlling the session
- Session participant: the session participant who is not the group controller
- Session controller: the logical resource controlling session
- VoIP bridge: the logical resource acting as VoIP bridge
- Chat server: the logical resource acting as chat server
- InitSes: the message for initiating a new session
- InitSesAck: the message for acknowledgement of creation of a new session
- XmitSpeech: the service event carrying speech from the participant
- RecvSpeech: the service event carrying speech to the participant
- XmitChat: the service event carrying chat message from the participant
- XmitChat: the service event carrying chat message to the participant.

We have not described dynamic views relating to the use case of Figure 8.2.

8.5 Link to Service Management

We shall next describe how the service model for our DiffServ example can be used in product, service, and resource management processes. We also make a few notes on how policies could be related to the example at hand.

8.5.1 Service Configuration

Taking into use of our example service requires some configurations to be made. These can be divided into two classes: overall ramp-up of the service on one hand, and adding a subscriber to the service on the other. We shall describe these next. We consider only operations related to Aggregate VoIP, and not ones dealing with, for example, DiffServ connectivity.

Service launch

The overall taking into use of the service involves multiple tasks. Here, we shall concentrate on service quality related ones that are related to the service provider and the connectivity provider.

The service provider uses service framework for describing the structure of the end-user service, and the service quality requirements of individual service events. The connectivity provider uses the service framework to communicate available service quality support aggregates (PHBs) and the relevant PDBs to the service provider. The service provider maps service events to traffic aggregates, and uses the service framework to convey mappings to the connectivity provider. Service provider also describes target end-to-end performance levels using service framework and provides a subset of this to the subscriber as part of the SLA.

The connectivity provider uses the mappings described using the service framework for configuring DiffServ edge routers. In the simplest case, all routers can be configured with a rule saying that the mapping rules in question must be applied to all the packets heading for or coming from the service provider's domain. More complex rules may be used in practice to support avoidance of Denial of Service (DoS) attacks, for example.

Adding a subscriber

In the simplest case described above, adding a new subscriber to the augmented VoIP service does not require any operations from the connectivity provider. This is true given that the subscriber already has an agreement with the connectivity provider and that all edge routers are configured with an 'uplink' rule relating to the mapping of traffic destined to the service provider domain.

8.5.2 Service Assurance

The information about the actual service level can be gathered by the subscriber and the service provider for end-to-end performance, and by the connectivity provider for performance within the transport domain. The service framework can be used for exchanging information between stakeholders, and for comparing target values to actual ones.

The service provider can make throughput measurements within the service domain by using active measurements (sending test traffic). The service provider can also measure service quality passively at selected test endpoints, and by utilising feedback from the Real-Time Control Protocol (RTCP) in media streams. These methods can be supplemented with active emulation measurements (cf. e.g. (Lakaniemi *et al.*, 2001)), where necessary.

The resources associated with services were not described above. With a full service topology model including resource linkage, it is possible to automate Key Performance Indicator (KPI) aggregation and assessment of performance degradations and failures on end-user services.

8.5.3 Service Portfolio Management

The example consists of a single service, and consequently the service portfolio management as such is not relevant to this case. The information about the usage of the service can be used as a basis for decisions relating to changing the parameters of the service. For example, the attractiveness of low-throughput data up/downloading could be enhanced by increasing the throughput.

The constituent product-oriented services can be reused as parts of other products, either as such or as templates.

8.5.4 Resource Development

Both the service provider and the connectivity provider can use anticipated and actual information about the service usage represented with the service framework for resource development purposes. For the service provider, such decisions relate to the service resources and the capacity of the LAN. The connectivity provider performs resource development relating to core network transport capacity on one hand and DiffServ edge routers on the other. The information about the trends in usage patterns in the temporal and geographical context are valuable for this purpose. Again, the full service model including resource linkage can be used for automating the process.

8.5.5 Product Management

The information about the usage of the service can be used for adjusting charging modes and rates relating to individual service events. The same information can also be used for determining changing the packaging of the end-user service, for example, inclusion into a larger product package. Where available, demographical usage information can be used in marketing.

The augmented VoIP can be used as the building block for constructing composite products.

8.5.6 Use of Policies

Provider type stakeholders can use policies to automate management tasks within respective domains. The connectivity provider can have default policies for most common service event types, for example, mapping VoIP media streams to EF PHB, browsing to AF11, and other traffic to BE PHB. This kind of rule package can be provided at a cheaper price than one requiring more customisation (and hence human participation). Indeed, as part of describing the use of the service framework above, the service aggregate level default of mapping all service events to AF22 was used.

8.6 Summary

Our first example modelled aspects of the DiffServ example, including the use of roles for specifying service-level definitions. In our example, service modelling can be directly used by the service provider and the connectivity provider. In order to be used optimally, the full service model including linkage of resources to service topology should be used. Such an environment makes it possible to utilise the service model for automating service assurance and resource development, for example.

8.7 Highlights

Ten things to remember from this chapter:

- The example relates to the DiffServ transport network.
- Classical DiffServ is based on static service quality provisioning.
- The uplink and downlink service quality provisioning are conceptually different for 1 : N provider/customer relations in DiffServ.
- Connectivity and the service provider are separate stakeholders in this example.
- Augmented VoIP is used as an example service.
- End-user class specific variants were illustrated in the example.
- Charging for augmented VoIP was illustrated.
- PDBs are used for describing the transport domain contribution to end-to-end service quality.
- Telephony is a demanding real-time service.
- Service quality support in LAN is beneficial for VoIP.

9

Mobile Network Example

Having flexed our modelling skills with the relatively simple DiffServ example, let us be more ambitious and describe a more challenging scenario.

9.1 Introduction

Our example involves multiple mobile network operators and a different kind of value net thinking than the previous example. The value net is in a state of change at the moment for reasons described in Part One, and our example seeks to capture a scenario which can be a result of a natural evolution from the traditional Third Generation Partnership Project (3GPP) mobile network environment.

As a technical platform, mobile network brings some new dimensions as compared to the previous example. Being a wide-area wireless access technology, it is usually available virtually anywhere and at any time. As such, it provides exciting new possibilities for innovative services. 3GPP mobile networks also provide advanced multi-service support in a complete architectural framework.

We shall describe the 3GPP-specific issues relatively briefly in the following text and refer the reader to Appendix A for more information.

9.2 Description

The participants in the example include the end-user, subscriber, physical mobile network operators, a Mobile Virtual Network Operator (MVNO), provider of IP Multimedia Subsystem (IMS) services, and external service providers. The physical mobile network acts as a connectivity provider and owns all the physical network elements. In essence, the Physical Mobile Network Operator (PMNO) sells mobile connectivity service to the MVNO. From the MVNO's viewpoint, it uses the physical mobile network for connectivity towards the end-users and – as a minimum – has only necessary subscriber registers such as the Home Location Register (HLR) or the Home Subscriber Server (HSS). It is assumed that the IMS services for the example in question are provided by a party separate from both PMNO and MVNO. Furthermore, it is assumed that MVNO subcontracts content services such as news and weather for its portal from external providers.

Service Modelling: Principles and Applications
Vilho Räisänen

We assume that the subscriber has an agreement towards MVNO only, that services are subscribed to via MVNO and that service usage is charged as a part of the mobile phone usage bill. We further assume that end-users, employees of the subscriber, can activate new services themselves and are able to manage service-specific preferences.

We consider three services in our example: presence services based on IMS, multi-player chess gaming service using IMS, and MVNO portal service having news service and weather as constituents of the portal service. It is assumed that presence service is available to all end-users, but news, chess, and weather need personal subscription.

9.2.1 Stakeholders

An overview of stakeholders was provided in the preceding text. Here we provide an analysis of the roles each stakeholder plays.

In addition to the obvious usage role, the end-user plays the role of a subscriber when activating new services. The subscriber handles a default set of descriptions for the MVNO.

The PMNO acts as a low-level connectivity provider for the present example. From the viewpoint of MVNO, the PMNO is a service provider. Since only MVNO has an agreement towards the physical network operator, the latter is not directly visible to other stakeholders. From the viewpoint of the subscriber and the service providers, MVNO provides connectivity to services.

The IMS provider is a service provider for the MVNO. As is the case with content providers, it provides subcontracted services for the MVNO, which are charged by the latter party. We assume that there are separate content providers for news and weather in this example. In summary, from the viewpoint of the customer, MVNO also plays the role of the service provider for end-user services.

9.2.2 System Description

Let us now study the arrangement described earlier, in greater detail. As with the Diff-Serv networking example, we shall discuss inter-stakeholder arrangements first and then proceed to discuss issues relevant to each stakeholder. In contrast to the first example, we use roles in the latter stage, whereby an individual stakeholder entity can appear in multiple classes.

Inter-stakeholder arrangements

It is assumed that provider-type stakeholders have formal agreements between them describing the services and also covering Service Level Agreements (SLAs). From the viewpoint of end-user services, MVNO has top-level SLAs towards other providers and customers and uses SLAs towards physical mobile network essentially as a resource. The agreement between the MVNO and the customer includes General Packet Radio Service (GPRS) and Universal Mobile Telephony System (UMTS) access and IMS-based presence by default and allows end-users to enable other services themselves.

Service provider

The IMS operator provides two end-user services: presence and chess. Presence means a system where users of mobiles can indicate their present availability and preferred means of communications to selected other users by using IMS. Thus, for example, Joe can perform a check on Jane's availability from the IMS and find out that she is currently in a meeting and can only receive an SMS and e-mail. The reachability support of IMS is also used by the chess game in our example for locating and contacting opponents. Joe can, for example, use Jane's e-mail address Jane@provider.com to contact her. The chess game itself can be handled with endpoint applications, possibly supported by an application server at the service provider's premises. We assume that presence is part of the monthly subscription fee, and session-based charging is applied to chess games.

Let us next study the content providers who are responsible for news and weather content. They provide content for the end-user service which is the MVNO mobile portal. The news service is most likely arranged into multiple categories such as domestic, international, business, and society news. It is assumed that unlimited amount of browsing of news titles is included into a monthly news service fee, and news footage video clips are charged on per-viewing basis. The weather service includes countrywide weather information, local weather forecast, and Doppler radar images about rainfall. Access to countrywide weather forecast could be assumed to be part of the news package, and local weather and Doppler radar may be charged per viewing.

The MVNO is a service provider of multiple roles in our example. Firstly, it provides cellular connectivity service between the end-users and the services and the Internet in general. In this role, MVNO has agreements towards both service providers and subscribers. MVNO also performs charging and billing on behalf of the IMS provider and content providers. Finally, MVNO acts as a service aggregator for the portal service, composing end-user service (portal) using news and weather as building blocks.

The PMNO acts as a service provider for MVNO.

Access provider

From the viewpoint of this example, MVNO is the sole customer-facing access provider. In this role, it needs to provide SLAs towards both provider parties and subscribers. The overall format that SLAs follow is most likely the common structure of fixed Internet access, but there are a couple of mobile network–specific issues here.

Firstly, geographical coverage is a specific issue in countries where indoor coverage is not common. (In Helsinki, Finland, we have had quite a decent cellular coverage in subways for many years.) Thus, SLA may have clauses relating to geographical applicability. Conversely, SLA may have a specific clause for SLA in the primary location of a major customer.

Secondly, UMTS and GPRS networks provide more advanced multi-service support capabilities than fixed networks. This is especially true for mobile terminals. Thus, SLA in our example could address background data transfer, browsing, and streaming separately.

At the concrete level, connectivity is of course implemented using the PMNO's network. Subsequently, the SLA between the PMNO and the MVNO needs to address the same kind of location-specific issues and multi-service support compared to the customer-facing SLAs discussed above.

Customer description

Owing to the complex inter-relationship structure, all stakeholders except the PMNO in our examples either have a customer role associated with them or are actual customers.

The subscriber is the most obvious customer and has an agreement with the MVNO relating to cellular connectivity and presence services for its end-users (employees). It is assumed that the customer orders handsets for the end-users and makes sure that they are configured correctly. The subscriber is interested in ensuring that the conditions of the SLA are fulfilled.

End-users play the role of a subscriber when they provision chess service, news service, or weather service for themselves. They are also the ultimate users of all end-user services and are the primary source for end-to-end service performance. For self-provisioned services, the link to the decision about continuing or discontinuing service subscription is more direct than for services provided by the employer.

The MVNO plays the role of the customer for subcontracted services (presence, news, and weather). MVNO is also a customer of the physical connectivity service provided by the PMNO.

The IMS provider and content providers are customers of the MVNO in view of the connectivity and charging that the latter party provides. The fact that they play this role does not necessarily lead to separate agreements, but most likely affects the terms of the contract between them and the MVNO.

9.3 Service Framework

Let us now describe the application of the service framework for our example. We shall use the same overall format as the DiffServ example and describe the application of the concept of service from the different viewpoints discussed in the preceding text.

As we shall see, service framework is used for information exchange between stakeholders in various customer–provider role pair configurations.

9.3.1 Aggregate Service

In the following paragraphs, we describe four different types of aggregate services: the physical mobile network service, the MVNO service towards subscribers, the MVNO service towards providers, and the provider service towards MVNOs. As discussed above, the two last ones are most likely handled in a single contract.

Physical Mobile Network Operator

This is the description of the service provided by the PMNO towards the MVNO. The terms related to revising the terms of the agreement are important in view of the ability of the MVNO to accommodate growth in traffic volumes.

- Business-oriented parameters
  - Geographical coverage of the service, taking into account different traffic classes
  - Validity period of the connectivity service

  - Means of supervising and reporting SLAs
  - Procedures related to revising the terms of the agreement
- Service quality requirements in geographical context
  - Availability
  - Traffic class–specific parameters
- Service quality characteristics in geographical context
  - Traffic usage patterns per traffic class.

MVNO towards subscribers

Here, we shall describe the services provided by the MVNO towards subscribers. Three services are relevant here: basic cellular connectivity, presence, and mobile portal. We assume only the single end-user class here.

Basic cellular connectivity:

- Business-oriented parameters
  - Geographical coverage of the service
  - Validity period of the end-user service
- Service quality requirements in geographical context
  - Availability
  - Traffic class–specific parameters
  - Throughput
- Service quality characteristics in geographical context
  - Usage statistics.

Presence:

- Business-oriented parameters
  - Validity period of the end-user service
- Service quality requirements
  - Availability
  - Service instantiation time
  - Response time for presence information queries
- Service quality characteristics
  - Usage statistics.

Please note that geographical availability is assumed to be determined by cellular bearer availability and, subsequently, there is no need to state this information in the context of the end-user services separately. Also note that this assumption is not always true in mobile networks, but the geographical availability of a service can be defined to be smaller than that of a mobile bearer.

Portal service:

- Business-oriented parameters
  - Validity period of the end-user service

- Service quality requirements
  - Availability
  - Response time for browsing
  - Throughput
- Service quality characteristics
  - Usage statistics.

MVNO towards providers

Services provided by the MVNO towards the providers are described in the following text. We shall handle connectivity and charging separately. Note that providers need connectivity service in our example only for those end-user services that are not subcontracted by the MVNO. For presence, connectivity service is not needed in our example set-up. (It is of course possible that the IMS provider also sells it separately by using a cellular operator as a pure access provider.)

Connectivity:

- Business-oriented parameters
  - Geographical coverage of the service
  - Traffic class–specific parameters
  - Validity period of the connectivity service
  - Means of supervising and reporting SLAs
  - Procedures related to revising the terms of the agreement
- Service quality requirements in geographical context
  - Traffic class–specific parameters
- Service quality characteristics in geographical context
  - Maximum traffic volumes per traffic class
  - Token bucket parameters.

Charging:

- Business-oriented parameters
  - Applicable services
  - Validity period
  - Monetary settlements
  - Procedures followed in revising the contract
- Service quality requirements
  - Charging accuracy
- Service quality characteristics
  - Maximum volume of services.

Providers towards MVNO

The providers have three different services with associated definitions towards the MVNO: news, weather, and presence. It may be useful to point out that information about the maximum service usage volume is an important aspect of the provider's resource strategy.

News:

- Business-oriented parameters
  - Validity period of the end-user service
  - Two variants: business and non-business
- Service quality requirements
  - Availability
  - Response time
- Service quality characteristics
  - Maximum service usage volume.

Weather:

- Business-oriented parameters
  - Geographical coverage of the service: entire access network
  - Validity period of the end-user service: until further notice
  - Two variants: business and non-business
- Service quality requirements
  - Availability
  - Response time
- Service quality characteristics
  - Maximum service usage volume.

Presence:

- Business-oriented parameters
  - Geographical coverage of the service: entire access network
  - Validity period of the end-user service: until further notice
  - Two variants: business and non-business
- Service quality requirements
  - Availability
  - Response time
- Service quality characteristics
  - Maximum service usage volume.

9.3.2 Service Variants

We shall not consider end-user class specific variants since we have covered the basic concept already in the DiffServ networking example and also in the interest of limiting the length and complexity of this example. We shall also leave out access technology– specific variants, even though they could be easily accommodated into the service framework. Because of this, we have one service variant per aggregate service.

We shall only consider variants relating to end-user services. Similar descriptions could easily be given for other services too. Please note that chess game is viewed to be an application using the IMS presence information.

Cellular connectivity

The parameters of the cellular connectivity are determined by the customer's Quality of Service (QoS) profile stored in HLR, as described previously. This information is provisioned per Access Point Name (APN) in Gateway GPRS Support Nodes (GGSNs), whereby different services may have different performance levels. We shall assume that there are three default APNs corresponding to background, interactive, and streaming traffic classes. Possible session-based services are assumed to use secondary APNs using 3GPP R5+ mechanisms.

The composition of cellular connectivity in our example in terms of service events is as follows:

- Background traffic
- Browsing
- Non-session-based streaming.

Parameters include the following:

- Business-oriented parameters
  - Availability
- Technical parameters
  - Service quality support is instantiated using UMTS and GPRS bearer activation or modification mechanisms
  - Charging per volume applied to traffic according to the traffic class
- Service quality requirements
  - Bearer activation
  - Traffic class–specific 3GPP QoS parameters
- Service quality characteristics
  - Usage information.

If we had session-based real-time services in our example, service quality support instantiation would require 3GPP PDF for authorisation.

Portal service

Portal service is assumed to consist of a multitude of different items, with only news and weather described here. It is assumed that both news and weather are accessed from the servers in respective providers' domains and are not 'mirrored' in the MVNO domain. Caching can still be applied locally, where necessary.

The composition in terms of service events is as follows:

- Portal home page access
- News service access: news browsing
- News service access: video clips
- Weather service access: countrywide weather
- Weather service access: local weather
- Weather service access: Doppler radar images.

Parameters include the following:

- Business-oriented parameters
  - Availability
- Technical parameters
  - Portal access requires MVNO Wireless Application Protocol (WAP) gateway
  - News access requires news provider WAP gateway
  - Weather access requires weather provider WAP gateway
  - Charging modes applied
  - Viewing of news clips in downloaded mode requires Digital Rights Management (DRM) functionality
  - Viewing of Doppler images in animated mode requires Java(tm) functionality in the terminal
- Service quality requirements
  - Service instantiation time (first access to portal)
  - Response time for browsing
  - Performance for streamed/downloaded video clips
- Service quality characteristics
  - Usage information stored separately for each service event type
  - Token bucket parameters for browsing and video clip access.

Presence service

Presence service includes management of access rights, updating one's presence information in the IMS, and accessing of others' corresponding information.

The composition in terms of service events is as follows:

- Access right management
- Updating of presence information
- Access to presence information
- Portal home page access.

Parameters include the following:

- Business-oriented parameters
  - Availability
- Technical parameters
  - All actions of presence service require the presence server in the IMS
  - All operations related to presence require that the user authenticates her/himself to the presence server. Automation of this may require conveying of identity such as the International Mobile Subscriber Identity (IMSI) between the MVNO and IMS operators
  - Charging modes applied
- Service quality requirements
  - Service instantiation time

  - Response time for access right update
  - Response time for presence information update
  - Response time for presence information access
- Service quality characteristics
  - Usage per service event type
  - Token bucket parameters.

Chess service

Chess service involves locating an opponent, game control signalling, and moves.

The composition in terms of service events is as follows:

- Opponent locating
- Start game message
- Pause game message
- Stop game message
- Move.

The parameters include the following:

- Business-oriented parameters
  - Availability
- Technical parameters
  - Locating an opponent requires presence service
  - All the other actions of the presence service require the chess application server in the IMS
  - Charging modes applied
- Service quality requirements
  - Service instantiation time
  - Performance target for game control messages
- Service quality characteristics
  - Usage per service event type
  - Token bucket parameters.

Note that the performance of the opponent location is determined by the performance of the presence service.

9.3.3 Service Events

No fewer than 17 different service events have been described so far. We shall not consider the service events relating to the cellular bearer here, leaving us with 14 service events: portal home page access, news browsing, news video clips, countrywide weather, local weather, Doppler radar images, presence access right management, updating of presence information, access to presence information, chess opponent locating, start chess game, pause chess game, stop chess game, and chess move. We shall go through them briefly in the following paragraphs.

Portal home page access

This service event consists of accessing the portal WAP home page using a browser in the handset.

- Service quality requirements
  - Service quality of designed type
  - End-to-end delay: interactive
  - Packet loss is relatively low to avoid retransmissions
- Service quality characteristics
  - Traffic patterns: temporally randomly generated events within a session.

News browsing

This service event amounts to accessing news content linked to the portal WAP home page using a browser in the handset.

- Service quality requirements
  - Service quality of designed type
  - End-to-end delay: interactive
  - Packet loss is relatively low to avoid retransmissions
- Service quality characteristics
  - Traffic patterns: temporally randomly positioned events, usually temporal correlation between browsing individual news items.

News video clips

This service event consists of viewing of news video clips with a handset. Video clips can be either downloaded for viewing or viewed using streaming.

- Service quality requirements
  - Service quality of designed type for clip downloading, inherent for streamed video
  - End-to-end delay interactive for initiating download or streaming
  - Packet loss is relatively low
- Service quality characteristics
  - Traffic patterns: clip access is temporally random, video clip request is followed by a large download event (accessing the clip)
  - For streamed clip viewing, the bearer with higher traffic class is needed.

Countrywide weather

The description of countrywide weather is assumed to consist of textual description as well as pictorial data. It is accessed with WAP.

- Service quality requirements
  - Service quality of designed type

  - End-to-end delay interactive
  - Packet loss is relatively low to avoid retransmissions
- Service quality characteristics
  - Traffic patterns: temporally randomly generated events within a session
  - Download events (responses) are larger in size than upload (request) ones.

Local weather

Local weather is assumed to be in the same format as the countrywide weather, and subsequently the same service event descriptions apply.

Doppler radar images

Doppler radar images are usually represented either as animated sequences or as a set of multiple images. In both the cases, the amount of image data is usually larger than in the case of weather maps.

- Service quality requirements
  - Service quality of designed type
  - End-to-end delay interactive
  - Packet loss is relatively low to avoid retransmissions
- Service quality characteristics
  - Traffic patterns: temporally randomly generated events. Downlink events are larger in size than uplink events.

Presence access right management

Presence access right management consists of viewing the present configuration, updating the message to the presence server, and acknowledging (positive or negative) the updating. Updating is performed using WAP browsing and has the same kind of service event quality requirements and characteristics as, for example, news browsing.

Presence information updating

Presence information update consists of the update message and acknowledgement. The requirements and characteristics are the same as for presence access rights management.

Presence information access

Presence information access consists of the access update message and reply. The requirements and characteristics are the same as for presence access rights management.

Chess game events

Earlier, we identified five different events associated with chess game: opponent locating, game control (start/pause/stop), and chess move. They can all be assumed to have roughly the same characteristics, consisting of small messages followed by a reply (opponent location) or acknowledgement (other events). As such, they have requirements similar to the presence service control related events.

9.3.4 Service Event Types

Service events in this example are assumed to be implemented using the default APNs of the MVNO for different traffic classes, that is, background, interactive, and non-session-based streaming. They can be thought to be associated with 3GPP traffic classes bearing the same names.

The aggregation criteria for each of the event types reads 'all traffic flowing via APN'. This means that the operator configures classification criteria according to which packet is assigned to one of the APNs. We shall consider the classification criteria later on in our modelling.

Background traffic

This service event type is used for non-interactive data transfer, which has lower priority than, for example, browsing. Background traffic class does not have throughput guarantees, but the maximum bit rate limit can be applied.

- Technical parameters
  - Aggregation criterion: all traffic flowing via APN
  - Traffic conditioning method: buffering or dropping
  - Maximum throughput can be large
- Service quality requirements
  - Map to background traffic class and use subscriber's HLR profile.

Interactive traffic

This service event type is used for traffic that is related to interactive services such as browsing. The interactive traffic class does not have throughput guarantees.

- Technical parameters
  - Aggregation criterion: all traffic flowing via APN
  - Traffic conditioning method: buffering or dropping
  - Limit maximum throughput to match the message size of interactive services
- Service quality requirements
  - Map to interactive traffic class, and use subscriber's HLR profile.

Non-session-based streaming

Non-session-based streaming supports guaranteed bit rate. The maximum bit rate can be configured to correspond to typical values required by mobile video codecs.

- Technical parameters
  - Aggregation criterion: all traffic flowing via APN
  - Traffic conditioning method: dropping

- Service quality requirements
  - Map to streaming traffic class if terminal supports it, otherwise use interactive traffic class. Apply subscriber's HLR profile.

9.3.5 Note

In the 3GPP standard, traffic classes and other bearer QoS parameters are used for defining service quality levels. The allowable service quality range is provisioned using the APN mechanism in GGSN as described previously. Subsequently, APN parameters form the basis for service quality provisioning also towards external network. Usually there are rules at the network edge describing mapping of traffic classes onto DiffServ Per-Hop Behaviour (PHBs) and/or Multi-Protocol Label Switching (MPLS) Label Switched Paths (LSPs). Consequently, we shall not describe separate service event classes for external networks, but use the 3GPP service event types as the basis.

Classification criteria for static service quality provisioning can have finer granularity than APN, but we shall not discuss this possibility here for simplicity. Some discussion about this can be found in (Koivukoski and Räisänen, 2005).

9.4 Service Model

We shall describe the service model for the mobile network example. As with the previous example, we shall provide a set of views rather than a complete model. Compared to the DiffServ example, we place more emphasis on modelling of charging and the types of services involved.

9.4.1 Use Case View

A use case view for the mobile networking example relating to the creation of services is shown in Figure 9.1. Only stakeholder types are presented in the figure in the interest of brevity, but the use case could be taken to role granularity as described in Section 3.6.3 and (Service Framework, 2004). For example, service definition phase could be analysed in terms of participation of business and technical roles within each stakeholder.

The description of the entities in Figure 9.1 is as follows:

- MVNO: set of roles belonging to MVNO
- ExtProvider: set of roles belonging to the service providers
- PMNO: set of roles belonging to the operator of the physical mobile network
- EndUser: role of end-user
- Service definition: definition of the composition of the service. Includes both business and technical parts
- Service implementation: technical implementation of the service, including both implementations of new configurations and reuse of existing ones
- Service deployment: the set of actions required to make necessary configurations in order for the service to work
- Service provisioning: the set of actions required to enable the service for an end-user. Can be done by the service provider or by end-user (self-provisioning).

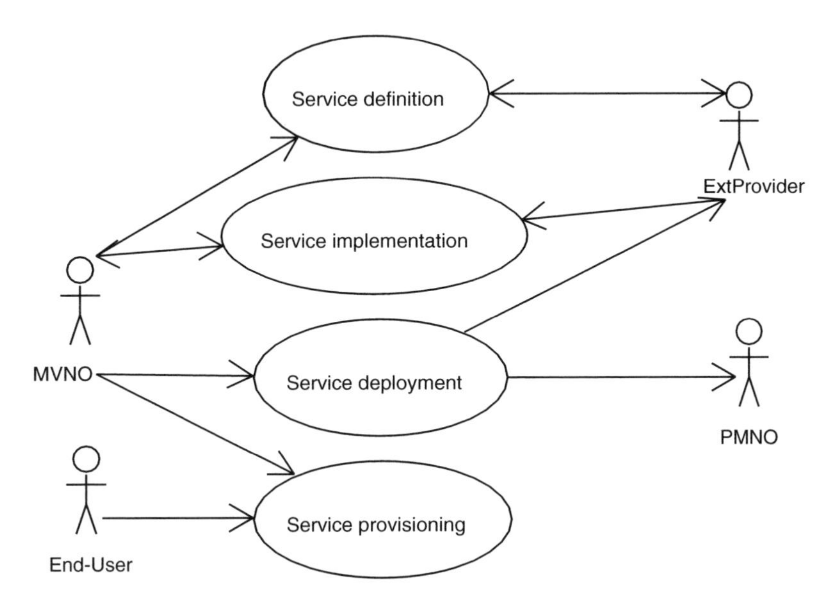

Figure 9.1 High-level use case for service creation for mobile networking example

Another use case – Figure 9.2 – represents use cases related to service quality definition and management for the portal service. It shows phases that are roughly parallel to the DiffServ network use case in Figure 8.2.

The description of the entities in Figure 9.2 is as follows:

- MVNO: as in the previous use case
- Service provider: as in the previous use case
- PMNO: as in the previous use case
- EndUser: as in the previous use case
- QoS profile definition: definition of service quality support for subscriber classes using 3GPP QoS profiles in HLR
- Service level definition: definition of the target service level of end-user services
- Service mapping: mapping of end-user services and constituent service events to service quality support classes
- Actual service level: collection of information about the actual service level
- 3GPP QoS management: management of the service quality support mechanisms using parameters.

In the above use case, MVNO is the service provider carrying the overall responsibility for the portal service and hence also for the related service quality. Other stakeholders make use of definitions created by MVNO. Aspects of measured service quality are also available to PMNO for the purpose of supervising fulfilment of SLA between MVNO and PMNO.

9.4.2 Static View

We shall begin the description of the static view of mobile networking example with a description of the inter-relations of end-user services and their relation to subscriptions.

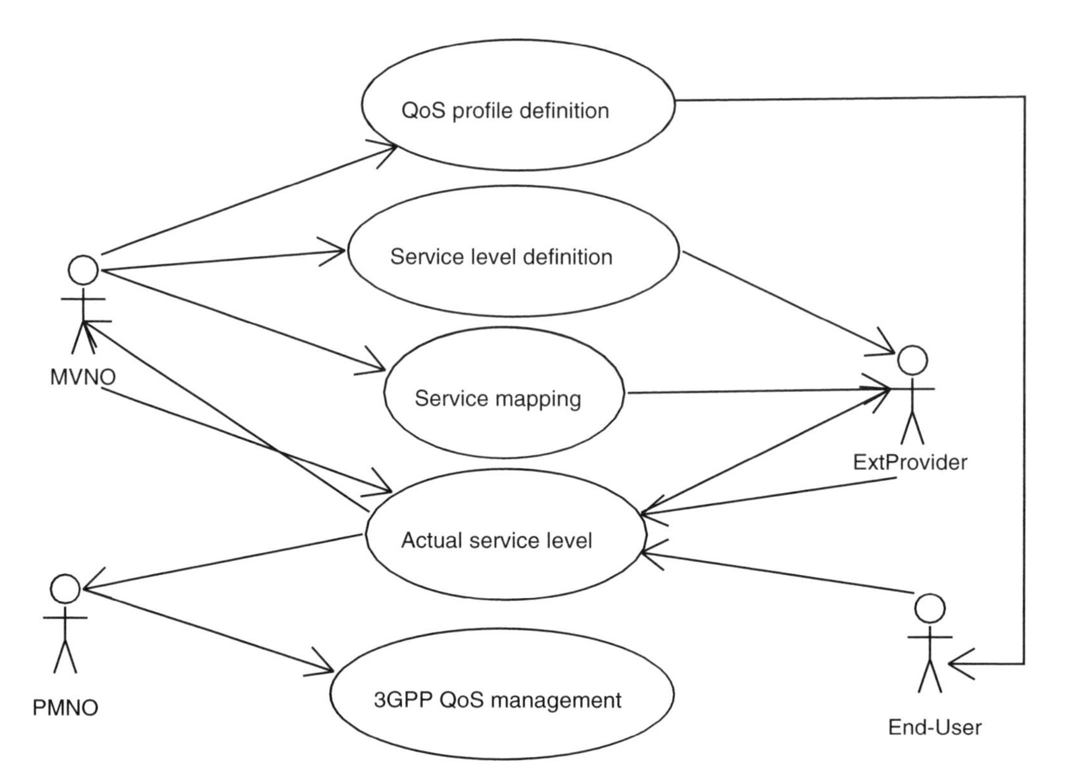

Figure 9.2 High-level use case for service quality management for the portal service within the mobile networking example

We shall deal with services in two parts. The first of these, the relation of end-user services, subscriptions, and providers belonging to the basic monthly package to each other, is shown in Figure 9.3. It illustrates the services that belong to the flat rate basic portal package described in an agreement between the MVNO and the subscriber. The provider parties responsible for constituent services are also indicated.

The description of the entities in Figure 9.3 is as follows:

- Subscriber: described previously
- MVNORole: MVNO's role as service provider
- PortalServiceFlatRate: the part of portal service covered by PortalAgreement
- PortalAgreement: Agreement between Subscriber and MVNORole concerning the use of PortalServiceFlatRate
- IMSRole: IMS provider's role in supporting end-user services
- Presence: Presence service
- PortalBrowsing: Portal browsing included in PortalAgreement.

Please note that the flat rate portal service has an aggregation relation towards both presence service and portal browsing. Using aggregation instead of inheritance allows also using these services separately from the flat rate portal browsing. Though it is not shown in the model above, MVNO could still apply templates for externally subcontracted

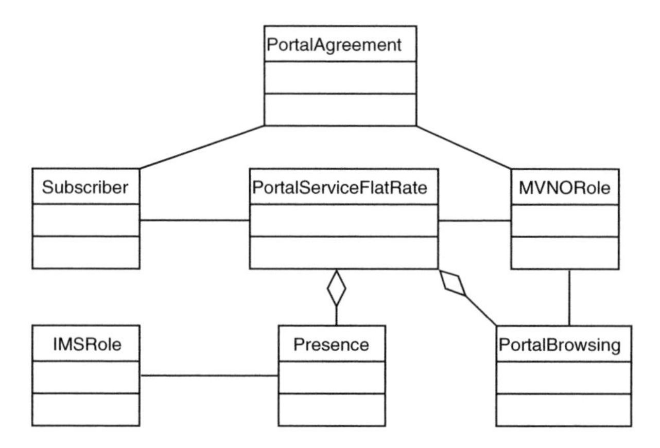

Figure 9.3 Inter-relations of flat rate end-user services and subscriptions

services. In such a case, the service provided as part of the portal on one hand and as part of other packaging on the other could be considered to be separate services.

The second part, relating to separately charged services, is shown in Figure 9.4. A convenience class, the generic portal service aggregates both flat rate and value-added portal services. An end-user, playing the subscriber's role, is responsible for agreements relating to value-added services.

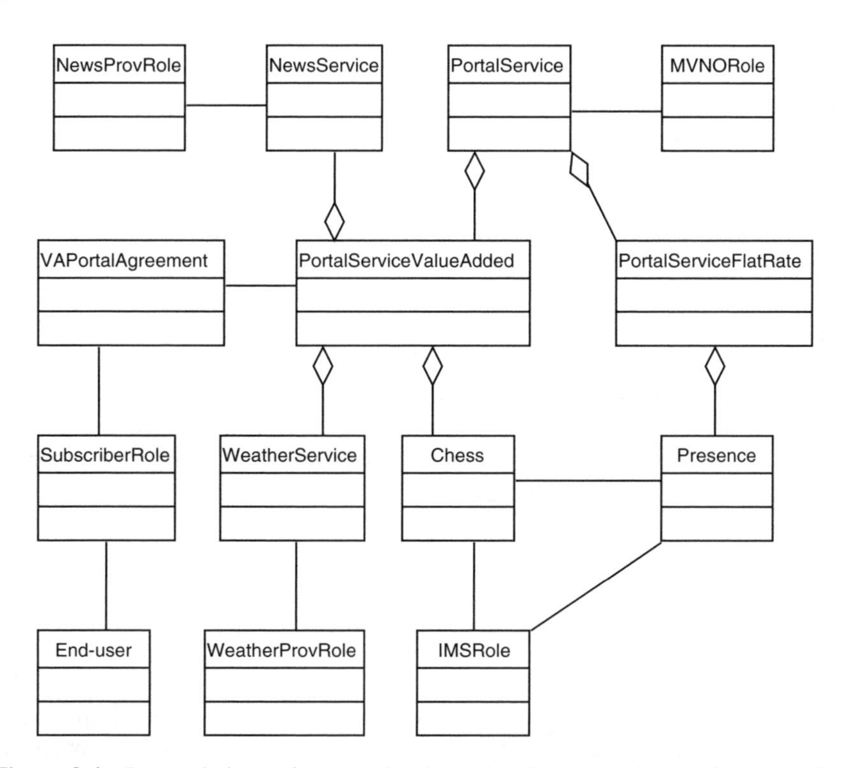

Figure 9.4 Inter-relations of separately charged end-user services and subscriptions

The description of the entities in Figure 9.4 is as follows:

- MVNORole: described previously
- IMSRole: described previously
- NewsProvRole: news provider as supporter of end-user services
- WeatherProvRole: weather provider as supporter of end-user services
- SubscriberRole: described previously
- EndUser: described previously
- PortalService: overall portal service provided by MVNORole
- PortalServiceValueAdded: value-added portal services
- PortalServiceFlatRate: described previously
- NewsService: news service
- WeatherService: weather service
- Chess: chess service
- Presence: describer previously
- VAPortalAgreement: agreement between SubscriberRole and MVNORole about the usage of value-added portal services.

Value-added portal services could depend on flat rate portal services, which is the case for chess and presence. This means that if presence were removed from flat rate portal services, it should be provided as part of value-added portal services, for example. The division into value-added and flat rate services relates to packaging of services by the MVNO, and subsequently an external service provider can provide both types simultaneously.

We shall next construct a model showing different kinds of services present in the system. As we have discussed previously, product-facing, abstract, and Resource-facing Services (RFS) can be viewed as having different sets of parameters and roles associated with them.

The service hierarchy will be shown in multiple parts for the sake of legibility. We shall model the chess service by way of an example. A similar approach can be used for other services.

The first picture (Figure 9.5) shows that presence and chess are both product-facing services, that chess depends on presence, and that both need an end-to-end 3GPP bearer to function. The end-to-end bearer is modelled as an abstract service. Please note that this figure does not contain information on the packaging of the products shown in Figures 9.3 and 9.4, but only shows the classification of the services involved with the chess product.

The description of the entities in Figure 9.5 is as follows:

- ProductFacingService: described previously
- AbstractService: described previously
- Presence, Chess: described previously
- E2EBearer: representation of end-to-end 3GPP bearer.

Modelling of an end-to-end bearer is shown in Figure 9.6. In this simple model, the end-to-end bearer depends on the Public Land Mobile Network (PLMN) bearer and the external bearer, which are assumed to be RFSs. The PLMN bearer includes everything

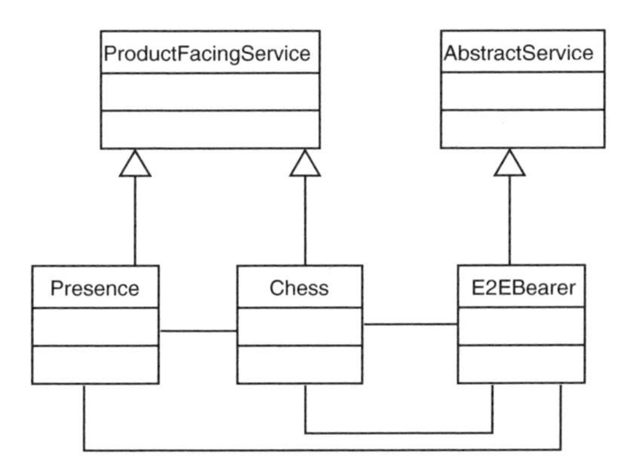

Figure 9.5 Top-level service example for the mobile network example

from communications endpoint up to GGSN and the external bearer includes the rest of the connectivity needed to reach the service. Note that the two bearers could have been modelled as abstract services, too, consisting of other services if a more detailed model would have been our goal.

The description of the entities in Figure 9.6 is as follows:

- AbstractService, ResourceFacingService: described previously
- E2EBearer: described previously
- PLMNBearer: bearer within 3GPP network
- ExtBearer: bearer outside the 3GPP network.

Next we shall drill down into relations of the PLMN bearers on the basis of the 3GPP standard architecture. We shall assume that either Wideband Code Division Multiple Access (WCDMA) or GPRS bearers can be used as part of an end-to-end bearer. It is a standard feature of WCDMA handsets at the moment; when third generation coverage is not available, the second-generation bearer is used. Note that this assumption implies that all end-users need to have WCDMA enabled for them. This is not the case at the moment, and second-generation–only subscribers could be modelled either separately or with the same model but always being out of WCDMA coverage.

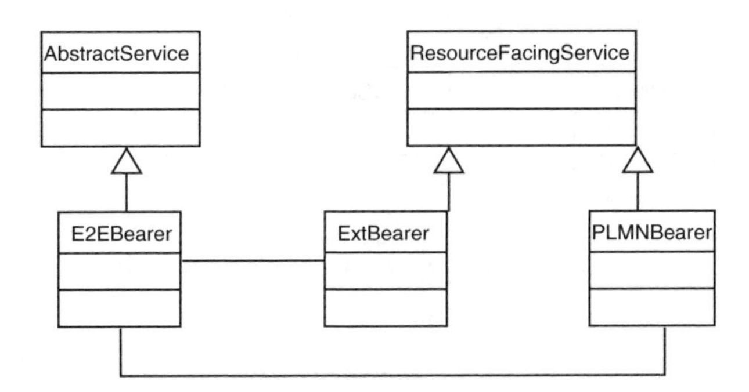

Figure 9.6 Dependencies of the end-to-end bearer for the mobile network example

In our simple model, the GPRS bearer requires the following resources to function: GGSN, Serving Gateway Support Node (SGSN), Base Station Controller (BSC), and Base Transceiver Station (BTS). The WCDMA bearer, on the other hand, needs the following resources: GGSN, SGSN, Radio Network Controller (RNC), and WCDMA base station (Node B). Earlier we noted that we could have made a more detailed model instead of the simpler option we chose. In an alternative model, some of the functionalities provided by GGSN, for example, could have been modelled as RFSs, whereas bearers would have been abstract services.

The description of the entities in Figure 9.7 is as follows:

- PLMNBearer: described previously
- GPRSBearer: PLMN bearer in GPRS network
- WCDMABearer: PLMN bearer in WCDMA network
- Resource: describer previously
- GGSN: Gateway GPRS Support Node of 3GPP network
- SGSN: Serving GPRS Support Node of 3GPP network
- BTS: Base Transceiver Station
- BSC: Base Station Controller
- RNC: Radio Network Controller
- NodeB: base station of WCDMA network.

Since the types of elements required by a bearer are indicated, the model of Figure 9.7 allows one to make availability checks for necessary resources. If, for example, no SGSN

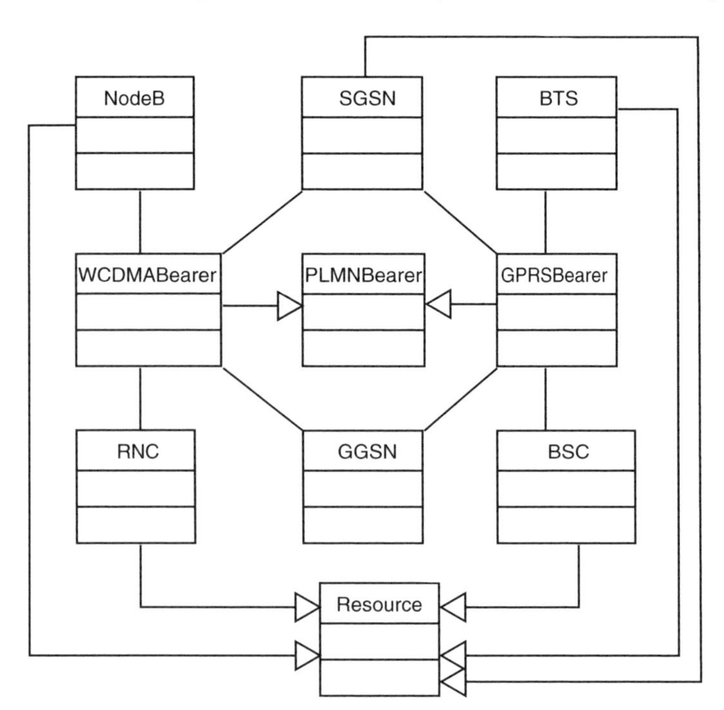

Figure 9.7 Types of bearers and their relations for the mobile network example

functions, neither the GPRS nor the WCDMA bearers work. Subsequently, end-to-end bearer and presence and chess services are also not available. Relation to particular SGSN is resolved as part of instantiation of the model for a particular use.

Above, GPRS bearer and WCDMA bearer were presented as RFSs used by the product-facing services. As 3GPP Internet connectivity can also be sold in a stand-alone fashion, Figure 9.8 illustrates the modelling of cellular data as a product. In this view, we retain the two RFSs and introduce the mobile connectivity product-facing service which is based on the PLMN bearer. The QoS profile corresponding to the end-user has been shown in the model, since it has a bearing on the service quality level reachable by the service. Note that, strictly speaking, the QoS profile relates to a subscription and should be associated with the end-user's role as a subscriber.

The description of the entities in Figure 9.8 is as follows:

- End-user: described previously
- PLMNBearer: described previously
- GPRSBearer, WCDMABearer: described previously
- MobileConn: mobile connectivity service
- APN: access point name associated with mobile connectivity service
- QoSProfile: end-users' QoS profile
- QoSProfileEntry: QoS profile entry for APN.

Next we shall consider service quality support provisioning for service event types. We shall assume here that no session-based QoS parameters are needed for streaming and that static APN-based provisioning is subsequently sufficient. Figure 9.9 shows a model for the use of APNs in service quality support provisioning. In the model, APNs and service event types have a one-to-one mapping between them. As shown in Figure 9.8, the service quality level achievable within a particular APN may be dependent on the user.

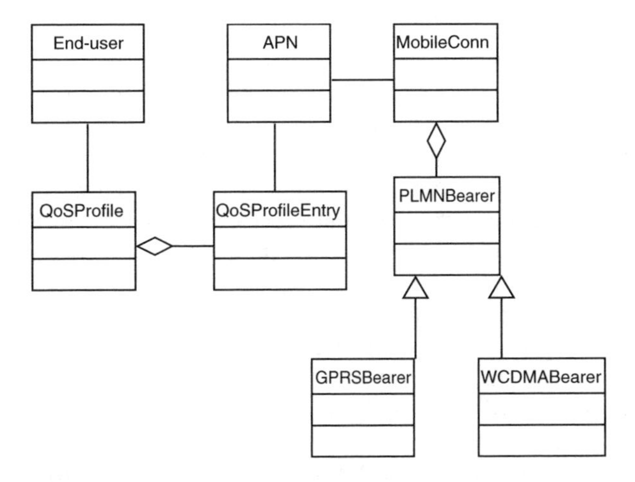

Figure 9.8 Modelling of the mobile bearer as a product

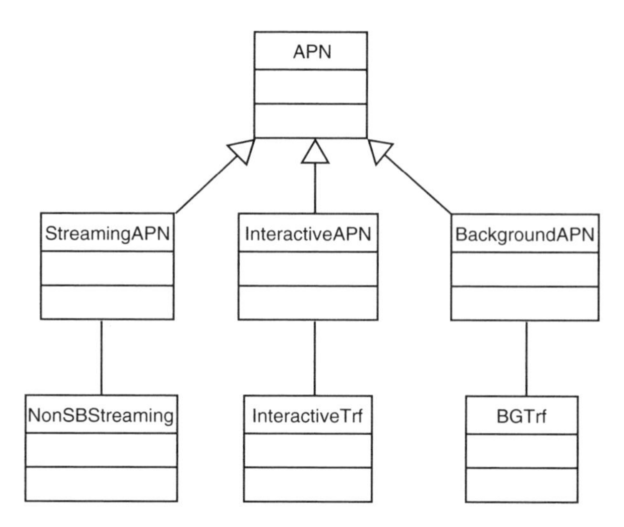

Figure 9.9 Service quality support provisioning

The description of the entities in Figure 9.9 is as follows:

- APN: described previously
- StreamingAPN: APN for non-session-based streaming
- InteractiveAPN: APN for interactive traffic
- BackgroundAPN: APN for background traffic
- NonSBStreaming: non-session-based streaming service event type
- InteractiveTrf: interactive service event type
- BGTrf: background service event type.

Let us take a closer look at how service quality is defined for services. Since all product-facing services in our example use the static service quality support provisioning paradigm, it is sufficient to study one of them. For dynamic – session-specific – service quality, the procedure is basically the same, except that the allowable service quality range corresponding to an APN is provisioned, instead of the maximum service quality level. The technical implementation involves the PDF in addition to the APN, as described previously.

We shall use streamed video clips as an example. Service quality support definition follows the pattern of Figure 7.25, so that service level definitions on an aggregate level define parameters such as availability which are specific to the aggregate level as well as parameters that are common for different service event types (in this case, browsing and streaming). We assume that APNs for both streaming and browsing are also used by service events belonging to other services.

The target end-to-end service quality for streaming is designed by the news service provider. The MVNO provides service level definition for the streaming APN, defining the minimum service quality level for streaming in mobile network domain Public Land Mobile Network, (PLMN). In addition to PLMN, terminal, transport network, and provider's own resources affect the end-to-end service quality. Assuming that transport part can be described with a DiffServ Per-Domain Behaviour (PDB), the model shown in Figure 9.10 can be drawn.

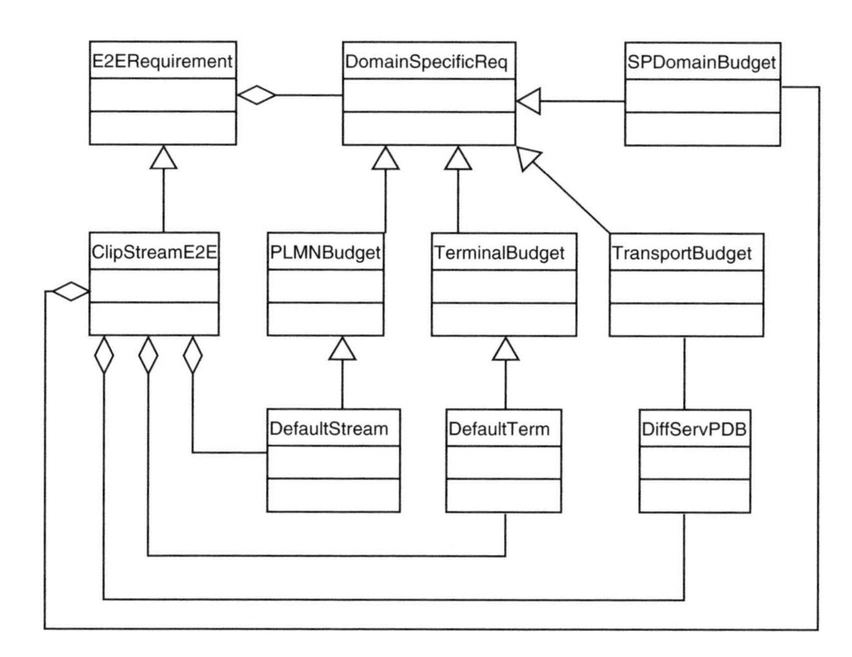

Figure 9.10 End-to-end service quality example for mobile network

The description of the entities in Figure 9.10 is as follows:

- E2ERequirement: described previously
- DomainSpecReq: described previously
- ClipStreamE2E: end-to-end requirement for video clip streaming, a type of E2ERequirement
- PLMNBudget: PLMN service level definition
- TerminalBudget: terminal effect on end-to-end service quality
- TransportBudget: effect of transport between service provider domain and PLMN on end-to-end service quality
- SPDomainBudget: effect of service provider's domain on end-to-end service quality
- DefaultStream: default streaming service level definition in PLMN
- DefaultTerm: default terminal service level impact
- DiffServPDB: DiffServ PDB relevant to PLMN service provider connection.

The default terminal impact on service level has been used above for simplicity. It should be noted that modern mobile networks have capabilities for discovering capabilities of individual mobile terminals, for adapting content to best match the terminals and for making configurations in terminals related to services.

End-to-end service quality can be further analysed in terms of stakeholders responsible for different factors contributing to it. A model for this is shown in Figure 9.11. In the model, it has been assumed that the PMNO is responsible for the transport network and hence also for DiffServ PDB. It is also assumed that the MVNO is able to provide a default service level impact definition for any terminal.

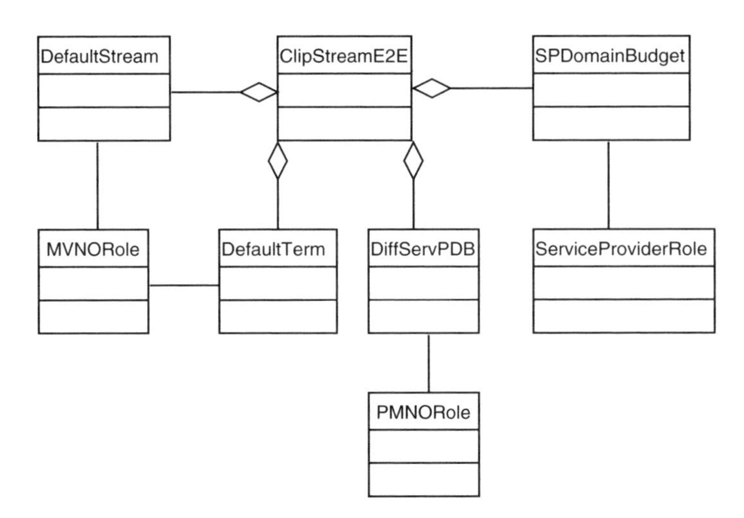

Figure 9.11 Relation of stakeholders to end-to-end service quality definition

The description of the entities in Figure 9.11 is as follows:

- ClipStreamE2E: described previously
- SPDomainBudget: described previously
- DefaultStream: described previously
- DefaultTerm: described previously
- DiffServPDB: described previously
- MVNORole, ServiceProviderRole: described previously
- PMNORole: physical mobile network operator.

As a final view to the static model, let us take a look at modelling of charging for the video clips. In the mobile network example, the charging mode to be used is defined by the service provider. Actual charging, on the other hand, is performed by the MVNO. Related modelling is shown in Figure 9.12 for the news services. A similar model can be drawn for flat rate portal services.

The description of the entities in Figure 9.12 is as follows:

- NewsService: described previously
- NewsProvRole: described previously
- NewsBrowsing: browsing-based news access
- NewsClip: viewing of streamed news clips
- PortalServiceValueAdded: described previously
- ChargingRole: set of tasks associated with charging
- MVNORole: described previously.

9.4.3 Dynamic View

We shall describe the self-provisioning of chess service as an example of the dynamic view. For our dynamic view, we take a step back and list the prerequisites for making

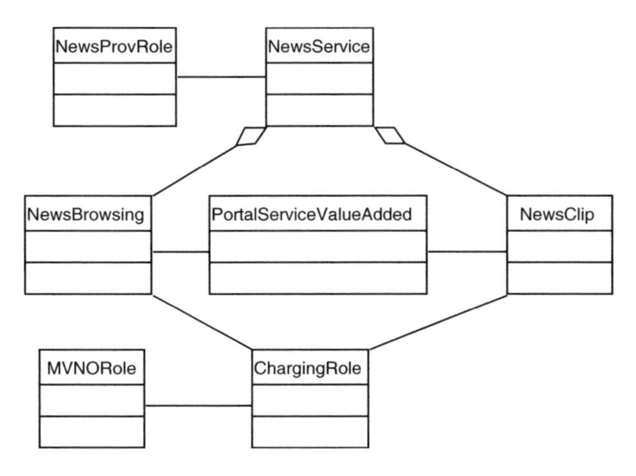

Figure 9.12 A model for charging in the mobile network

the service work, in addition to the actual self-provisioning operation. This kind of a 'big picture' is useful for troubleshooting, including customer interface operations such as help desks.

We illustrate the following phases: subscription activation, acquiring of communications settings, connection activation, and the actual provisioning of the service in Figure 9.13.

The description of the entities and messages in Figure 9.13 is as follows:

- End-user: described previously
- Subscriber provisioning: MVNO subscriber provisioning

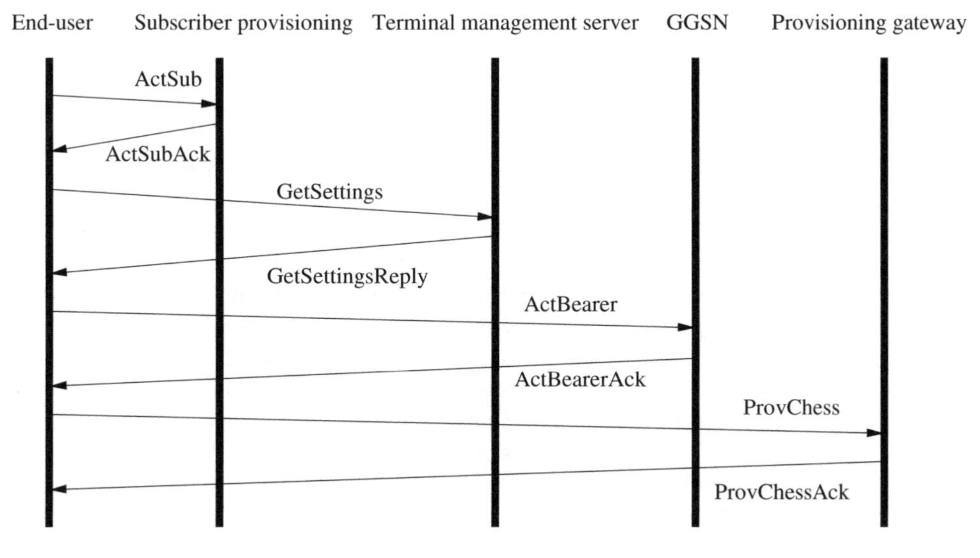

Figure 9.13 A dynamic view example for mobile network

- Terminal management server: server capable of detecting the terminal type and provisioning the necessary settings
- GGSN: described previously
- Provisioning gateway: portal for service self-provisioning
- ActSub: activate subscription
- ActSubAck: activate subscription acknowledgment
- GetSetting: request settings related to APNs, IMS, etc.
- GetSettingReply: relevant parameters returned
- ActBearer: bearer activation request
- ActBearerAck: bearer activation acknowledgement
- ProvChess: chess service self-provisioning request
- ProvChessAck: acknowledgment of chess service provisioning.

The communication settings given above relate to the names of APNs used for packet-based communication in 3GPP networks.

Service has been depicted as 'stand-alone' above, but obviously it could also be invoked from the portal service. The sequence diagram above is devised from the end-user viewpoint. Analogous sequences can be described for provider-internal activities, for example, configuration and provisioning of services.

9.5 Link to Service Management

We shall now discuss the link to service management. Compared to the DiffServ networking example, the participation of multiple stakeholders in service management is an essential part of our current example.

9.5.1 Service Configuration

Service configuration–related activities can be divided into product oriented, abstract, and resource service configuration. We shall discuss these in the following text. The classification into the above three categories has been made from the viewpoint of the mobile subscriber. It is useful to keep in mind that, for example, MVNO is a customer of the news service provider and the PMNO. Thus, the type of a service could depend on the role played by a stakeholder.

Product-oriented service configuration activities are operated by all provider-type stakeholders. MVNO operates portal services and Internet connectivity services, and the IMS provider operates presence and chess services. The News provider operates news service, which consists of browsing and video clip download services. Each of the product-facing services need to link to saleable products, of which we have studied the following three: operator portal service, news service, and chess service. The operator portal is a product package, and the constituent product-oriented services can be used as components in other products too. News service and chess service are products in their own right. The IMS provider could also sell presence directly as a separate product. In addition, Internet connectivity can probably be sold separately by the MVNO.

In modelling, we have considered only one abstract service, namely end-to-end bearer operated by MVNO. A real mobile operator could use abstract services for streamlining service management processes. The commonly used combinations of RFSs and abstract services could be represented as an abstract service.

We have considered two RFSs, namely, the PLMN bearer and the external bearer. We can associate these with MVNO and PMNO, respectively. In reality, a larger number of RFSs would be run by both the parties. For example, charging could be viewed as an RFS operated by MVNO.

Service configuration processes within each stakeholder make use of existing service topology information, update it as new service functionalities are created or modified, and keep information about product linkage up-to-date. Topology information describes the three kinds of services as well as their inter-relations.

9.5.2 Service Assurance

In our example, service assurance processes can be related to different kinds of services operated by stakeholders. Indeed, one of the strengths of moving towards service-oriented paradigm is the ability to identify entities as services to associate service level definitions to. Service topology information, together with service level definition for RFSs, can be used in determining and defining service levels for abstract services and, further, for product-oriented services. The service level definitions can then be used for triggering warnings and alarms related to different kinds of services and resources. Similarly, service topology can be utilised for assessing the impact of failures to service hierarchy and products.

Measurements and reporting form an essential part of the SLAs between different actors. In our case, the most important SLA interfaces are between MVNO and PMNO on one hand and between MVNO and service providers on the other. MVNO also has some kind of SLA towards the subscriber. When monitored entities are modelled as services, link between service assurance and modelling is natural.

In parallel with other kinds of wide area networks, service assurance information may be needed in geographical context. For example, MVNOs may have a major customer at a specific location and may require specific service assurance information for that location. Mobility support within one radio access technology, between technologies (e.g. GPRS and WCDMA) and also potentially between cellular and complementary access technologies such as 802.11 and 802.16 may be part of the specification.

9.5.3 Service Portfolio Management

Each of the stakeholders is assumed to be operating a repository of products as well as services of different types. The adding of portal service into the MVNO portfolio is accompanied by the addition of constituent services. Service portfolio information, including linkage to product and resources, can be assumed to be stored in a repository. For subcontracted services, the repository stores information about responsible stakeholders and interfacing information towards external services. The inventory may include an interface towards external providers for self-management of information related to services they are responsible for. Such an external interface may also support automated configuration and provisioning of new services.

The creation of new product-facing services also potentially requires the addition of new abstract and RFSs, or modifications to existing ones. New RFSs are required when new kinds of resources are brought into the system. For example, adding a chess game server into the IMS domain could be modelled as being accompanied by the addition of RFSs supporting chess service. In the management of this enabling services layer, abstract services can be of value, packaging RFSs into reusable sets of functionalities. New services can then be instantiated from a template or built on top of the functionality described in templates.

9.5.4 Resource Development

Resource development activities can directly make use of service topology information. Together with information about SLAs, service usage information, and service assurance, usage information can be aggregated and trends in service and resource usage can be identified and associated with types of resources. In the simplest case, the interface between MVNO and PMNO can be based on a small number of traffic aggregates – for example, 3GPP traffic classes – without making references to higher-level services. Between MVNO and the service providers, on the other hand, many kinds of arrangements are possible. Standard 3GPP APNs make it possible to provide service quality at a finer granularity than traffic class.

Practical tasks relating to resource development include ensuring sufficient service execution environment resources and connectivity towards external networks (external providers), ensuring that SLA towards PMNO can accommodate future traffic volumes (MVNO), and making sure that traffic aggregates can be supported in the whole coverage area according to SLAs.

9.5.5 Product Management

In our example, product management activities relate to portal product. The MVNO portal is a product packaging, making use of other products (3GPP bearer, presence, chess, news). Product management systems need to support involvement of business roles in assessing the feasibility of a new product, designing a solution to support successful designs, and formulating requirements for services to implement it. For the MVNO in our example, feasibility assessment involves reviewing partnering agreements towards external providers. The solution design phase takes into account SLAs towards external parties. For the 'upward flow', product management needs to be able to import information about product usage. Product management systems need to support the entire life cycle of products, including modification and retirement of products.

9.5.6 Use of Policies

It is not difficult to invent examples of the use of policies in our example. They could be used by all stakeholders. In our example, MVNO plays a central role, and, subsequently, we use it for illustrating the use of policies in our example.

One use of policies is providing a template for interfacing towards external actors. Basic connectivity packages for service providers can make use of policies, so that connectivity packages whose instantiation does not require changes can be provided at a lower rate than those that need customisation. In the latter case, provider-level policies can be overridden with a provider-specific, smaller-scope one.

9.6 Summary

We have modelled the aspects of a relatively complex value chain related to mobile networks. Representation of capabilities as services and using service modelling for conveying information about inter-relations is a powerful tool both within individual stakeholders as well as between them. In detailed modelling of the inter-relations, roles could be expected to play an important role. The same entity could appear in different roles for different stakeholders.

In our example, we have considered a model in which the mobile network operator subcontracts services from other providers and carries the overall responsibility for products and end-user services. This is by no means the only model, but the mobile network operator can also play the role of access facilitator for the service provider's own products. This scheme is closer to our first example, but certain differences still remain. It is typically most convenient to have the mobile network provider handle charging and billing of services.

9.7 Highlights

Ten things to remember from this chapter:

- Static provisioning mode was used for this example.
- Service quality provisioning was based on associating services with APNs.
- Non-session-based streaming was used in the example for the sake of simplicity.
- One APN per traffic class involved (background, interactive, and non-session-based streaming) was used in the example.
- MVNO is a re-seller for connectivity capacity contracted from PMNO.
- Two charging models were used in the example: subcontracted services belonging to the flat rate portal service and separately charged value-added services.
- Charging is always performed by the MVNO.
- MVNO provides charging and connectivity as services towards service providers.
- The effect of the terminal needs to be taken into account in determining the end-to-end service quality.
- In the sense of usage sequence, service usage is dependent on bearer availability.
- Availability of settings affects bearer availability.

10

Distributed Network Example

As our last example, we shall describe a future-oriented, distributed service, which involves peer-to-peer connectivity, managed connectivity, utility computing, and service composition. The example is not based on any particular architecture, but makes use of some of the ideas described in connection with MobiLife project in Part One.

10.1 Introduction

The service we shall consider here is different from the previous examples in that the overall responsibility for the service is with an end-user. Dynamic composition of the service is used, and automated negotiation about the price is part of the scenario. The service makes use of *ad hoc* group communications, and combines managed and non-managed service functionalities.

10.2 Description

The imaginary service in the example is peer-to-peer, *ad hoc* group service, which makes use of service functionalities via wide-area access. The service is assumed to be collaborative data analysis based on information gathered by participants and requiring significant processing power for analysis. Analysis of the collected data is performed remotely using a computing facility. Note that participants could also be easily and geographically distributed without major changes to the scenario.

10.2.1 Stakeholders

The stakeholders of the example are as follows:

- End-user 1
- End-user 2
- End-user 3
- Context broker
- Service provider
- Wide-area connectivity provider.

We shall next go through the stakeholders and discuss the kinds of roles they play in this example.

End-user 1 is assumed to be the coordinator of the group service, acting also as service provider for the data collection and processing service to other end-users. End-user 1 also interfaces to information processing provider for performing the actual data analysis. End-user 1 is a customer to wide-area communications provider. In order to locate information processing provider, the role of context information user is needed, too. Finally, end-user 1 also plays the role of connectivity provider (perhaps 'co-provider' is the most adequate term for a participant to *ad hoc* communications).

End-users 2 and 3 act as information providers for the information collection service, and participate in forming the *ad hoc* communications, too. They also act as users of the data collection and processing service.

Context broker provides, in our example, information about available services, making it possible to locate the information processing required by data analysis service.

10.2.2 System Description

For the described scenario to work, certain groundwork is needed, which is described in the following text.

The communications endpoints used by the participants to the data collection and analysis services are assumed to be able to gather data and attach necessary metadata to it so that it can be combined and processed meaningfully. Associated metadata can be based on specialised ontology only known to the relevant applications, or use a general ontology available in the network. In the latter case, we assume that ontology is already cached and does not need to be fetched for our example. Communication endpoints used by users 2 and 3 need to have means of conveying collected data to user 1. In our example, we assume that this is part of group communications functionality, either as part of the group communications support or terminal platform.

The endpoint operated by user 1 needs to be able to first discover the endpoints of users 2 and 3 using *ad hoc* communications and to make sure that necessary privacy and trust actions are invoked for the communication. We assume that user 1 is also the coordinator of group communication, and in this role manages group communication session. The data collection and analysis application run by user 1 need to be able to discover and import data provided by endpoints of users 2 and 3. The application in question can make use of metadata attached to measurement data for combining it with data from user 1 and for defining the computation task to be executed remotely. It is assumed that data analysis task can be described as a generic problem, so that any utility computing facility that understands the algorithm description can be used. For negotiating the price of the computation task, an interface is needed. Finally, user 1 needs to manage the overall choreography of the service.

Inter-stakeholder arrangements

Users 1, 2, and 3 need to have a trust relation between them and use compatible applications for sharing the fruits of data collection and analysis. User 1 is assumed to have an agreement with a connectivity provider, including access to context provider's facilities for locating services.

Service provider is assumed to have an agreement with context provider about making its services visible in the service repository. Thus, user 1 does not need to have a separate agreement with context provider.

Service provider

The computing service provider in our case provides utility computing service. It is assumed that service provider gives price information, which makes it possible for the use of the service to compare individual providers. Computing service is a managed service.

The context provider is assumed to host a repository of different kinds of service providers and their capabilities. It is assumed that necessary means for users to perform queries to repository exist. For our (relatively simple) example, we assume that exact matching can be used for repository access. This requires an ontology containing utility computing. Context access is a managed service.

User 1 provides data analysis service to users 2 and 3. It is a non-managed service.

Users 1, 2, and 3 provide *ad hoc* connectivity service to each other. It is a non-managed service.

Access provider

Connectivity provider operates Internet connectivity, and can be based on any access technology. In the interest of 'anywhere, anytime' analysis, we assume that cellular connectivity is used.

Users 1, 2, and 3 act as connectivity providers to each other.

10.2.3 Customer Description

Users 2 and 3 are customers of user 1. User 1, in turn, is customer of connectivity provider, context provider, and utility computing provider. The customer relationship among users is based on them using compatible application and executing the same task. The *ad hoc* group service makes use of the agreements of user 1 in the form of service functionalities for the overall service.

There is no formal agreement between users 1, 2, and 3 relating to data analysis service. Conceptually, data analysis can, nevertheless, be viewed to form a product; the use of which does not require monetary compensation.

10.3 Service Framework

We shall now proceed to describing application of the service framework to our example. Service framework is defined by user 1 – be it explicit or implicit – and used by all participants of the scenario in one form or another.

10.3.1 Aggregate Service

The top-level service in the example is data analysis service provided by user 1. It makes use of *ad hoc* group services, service discovery capability, and utility computing.

- Business oriented parameters
  - Coverage of the service: *ad hoc* peer-to-peer communications
  - Business parameters: provided 'as is' without any guarantees
- Service quality requirements: interactive responsiveness
  - Availability: no guarantees
  - Traffic class specific parameters
- Service quality characteristics
  - Information about connectivity performance gathered.

The information collected about connectivity performance can be used as a basis for optimising communications. Actual data gathering can be performed by data analysis application, for example.

10.3.2 Service Variants

Two variants of the service can be identified:

- Data analysis controller variant
- Data analysis contributor variant.

User 1 instantiates the data analysis controller variant, coordinating the collection and analysis of data. Users 2 and 3 instantiate data analysis contributor variant, participating in data collection and sharing of the results.

Controller variant

Composition in terms of service events is as follows:

- Group communications advertisement
- Service provider search message
- Price negotiation with utility computing provider
- Submission of computation task
- Returning of computation task results
- Sharing of the results of analysis within the group.

Parameters include:

- Business oriented parameters
  - Access technology: *ad hoc* (for group communications), cellular (for wide-area communications)
- Technical parameters
  - Application used for data analysis

- Service quality requirements
  - Not applicable
- Service quality characteristics
  - Not relevant.

Contributor variant

Composition in terms of service events is as follows:

- Group communications join message
- Data uploading from contributor to controller.

Parameters include:

- Business oriented parameters
  - Access technology: *ad hoc*
- Technical parameters
  - Application used for conveying data to controller
  - Data formats used
- Service quality requirements
  - Not applicable
- Service quality characteristics
  - Not relevant.

10.3.3 Service Events

The following service events can be identified:

- Group communications advertisement
- Group communications join message
- Data uploading from contributor to controller
- Service provider search from context provider
- Price negotiation with utility computing provider
- Submission of computation task
- Returning of computation task results
- Sharing of the results of analysis within the group
- Group communications termination message.

We do not consider possible messages related to setting up of *ad hoc* communications in our list.

We shall describe these briefly in the following text.

Group communications advertisement

This service event is broadcast in the *ad hoc* group to advertise the availability of group communications.

- Service quality requirements
  - Service quality of designed type

  - End-to-end delay: interactive
  - Packet loss relatively low to avoid retransmissions
- Service quality characteristics
  - Traffic patterns: temporally random one-off event per group session
  - Service event small in size.

Group communications join message

Participants to group communication issue this service event to indicate willingness to join. Acknowledgement of joining from group controller is considered to be part of the same service event.

- Service quality requirements
  - Service quality of designed type
  - End-to-end delay: interactive
  - Packet loss relatively low to avoid retransmissions
- Service quality characteristics
  - Traffic patterns: typically, temporally correlated with group communications advertisement
  - Service event small in size.

Data uploading from contributor to controller

This service event consists of transferring data from contributor to controller. Depending on the situation, data amount may be non-insignificant.

- Service quality requirements
  - Service quality of designed type, throughput preferably large
  - End-to-end delay: interactive
  - Packet loss relatively low to avoid retransmissions
- Service quality characteristics
  - Traffic patterns: temporally random
  - Event size may be large.

Service provider search from context provider

Service provider search event consists of description of the service sought for.

- Service quality requirements
  - Service quality of designed type
  - End-to-end delay: interactive
  - Packet loss relatively low to avoid retransmissions
- Service quality characteristics
  - Traffic patterns: temporally random
  - Service event small in size.

Price negotiation with utility computing provider

Price negotiation event consists of description of the computation task, and provider's reply indicating the price associated with the task.

- Service quality requirements
  - Service quality of designed type
  - End-to-end delay: interactive
  - Packet loss relatively low to avoid retransmissions
- Service quality characteristics
  - Traffic patterns: temporally random
  - Service event small in size.

Submission of computation task

This event includes the description of the computation task to be carried out, as well as associated input data. The latter may be large in size, for example, in the case of computation involving matrices or long time series vectors.

- Service quality requirements
  - Service quality of designed type, throughput preferably large
  - End-to-end delay: interactive
  - Packet loss relatively low to avoid retransmissions
- Service quality characteristics
  - Traffic patterns: typically, temporally correlated with price negotiation event
  - Service event size may be large.

Returning of computation task results

Results of the computation task are returned in this service event. Again, the amount of data included may be large.

- Service quality requirements
  - Service quality of designed type, throughput preferably large
  - End-to-end delay: interactive
  - Packet loss relatively low to avoid retransmissions
- Service quality characteristics
  - Traffic patterns: forms a pair with computation task submission, not necessarily temporally close
  - Event size may be large.

Sharing of the results of analysis within the group

The results of data analysis by user 1 are distributed to users 2 and 3 with this service event.

- Service quality requirements
  - Service quality of designed type, throughput preferably large

  - End-to-end delay: interactive
  - Packet loss relatively low to avoid retransmissions
- Service quality characteristics
  - Traffic patterns: typically, temporally correlated with returned computation results
  - Event size typically small.

Group communications termination

This service event is used to terminate group communications session.

- Service quality requirements
  - Service quality of designed type
  - End-to-end delay: interactive
  - Packet loss relatively low to avoid retransmissions
- Service quality characteristics
  - Traffic patterns: temporally random one-off event per group session
  - Service event small in size.

10.3.4 Service Event Types

In our example, all the service events are of interactive type. Even though the completion of a transaction may take some time, it is initiated in an interactive manner. Thus, interactive service event type is suitable for all of them. It is, nevertheless, useful to make a distinction between interactive events in the *ad hoc* network on one hand, and in the wide-area (cellular) network on the other. We shall describe these in the following.

Ad hoc interactive service event

There are different variants of *ad hoc* networks existing. Apart from classification according to the underlying wireless communication layer, a major distinction can be made between single-segment and multi-hop *ad hoc* networks. In the former one, each participant to the communications can directly contact all the others, whereas in the latter case this is not necessarily the case. In the latter case, individual communications nodes can route traffic between the other nodes.

One of the hallmarks of multi-hop *ad hoc* networks, in general, is the absence of guarantees about service quality. The routing topology may change at any time because of node mobility, and individual links between nodes may be severed and replaced by other links in a dynamical fashion. Even for single-segment networks such as a 802.11 infrastructure mode segment, no guarantees can be made owing to the reason that communications endpoints may move out of coverage, and other equipment operating on the same non-licensed frequency may reduce effective throughput and delay.

What is achievable in *ad hoc* networks is prioritised scheduling for interactive traffic in transmitting and/or relaying nodes as compared to treatment provided for data transfer traffic.

- Technical parameters
  - Aggregation criterion: all service events belonging to the data analysis service in *ad hoc* network
  - Traffic conditioning method: not relevant

- Service quality requirements
  - Map to the second lowest scheduling priority class on link layer.

In the preceding text, it is assumed that the lowest scheduling priority class is used for background data transfer traffic. Traffic conditioning is not applied, whereby our example application can use all available capacity for information transfer, unless pre-empted by a concurrent higher priority *ad hoc* service.

Wide-area interactive service event

Wide-area networks exemplified by cellular networks operate on licensed frequencies and are subsequently protected from interference by law. Owing to the cost of licensed spectrum, it is also in the interests of the connectivity providers operating on these frequencies to make the best use of the spectral resources. Subsequently, access networks such as General Packet Radio Service (GPRS) and Wideband Code Division Multiple Access (WCDMA) provide relatively advanced multi-service support. We have discussed the Third Generation Partnership Project (3GPP) Quality of Service (QoS) model previously, and have seen that it has a traffic class dedicated for interactive traffic.

- Technical parameters
  - Aggregation criterion: all traffic flowing via interactive traffic class APN
  - Traffic conditioning method: buffering or dropping
- Service quality requirements
  - Map to interactive traffic class and use subscriber's Home Location Register (HLR) profile.

Support of the interactive traffic class described above is based on the Access point name (APN) / HLR QoS profile-based provisioning method described in our previous example. The application employed by the controller variant of the data analysis service of user 1 is assumed to be able to request interactive traffic class bearer for the service.

10.3.5 Note

Our example shows how managed and non-managed entities can be used as part of a single end-user service. Both connectivity and service functionalities play these dual roles in our case. It is important to keep in mind that even though we may use concepts bearing the same title in managed and non-managed domains – discussion about service event types is a case in point here – the technical content of the entities may be different in each of the two domains.

10.4 Service Model

It is time to study modelling views of our example service. Following the pattern of the earlier examples, we put most emphasis on aspects not demonstrated in our previous models.

10.4.1 Use Case View

A subset of use case views to our service from usage point of view is shown in Figure 10.1. The use cases listed have been described earlier. We do not show uploading input data and transfer of the results of the computations in this view for the sake of brevity. Please note that users 2 and 3 play symmetrical roles, and are shown as a single actor below.

Description of the entities and messages in Figure 10.1:

- Create session: create group service session
- Upload data to controller: transfer data from contributor to controller
- Discover computing provider: locate provider of utility computing services
- Negotiate price for computing: negotiate price for computation task at hand
- Perform computation: perform computing task
- Distribute analysis results: distribute analysis results to contributors.

We shall not consider use cases related to creation of the service here. One reason for this is the fact that the managed components of the example are standard ones, so that tasks related to configurations made within context provider and computing provider are not specific to this example. Some of the issues relevant to wide-area networking were described within the previous example. Also, session termination is not depicted in the use case view.

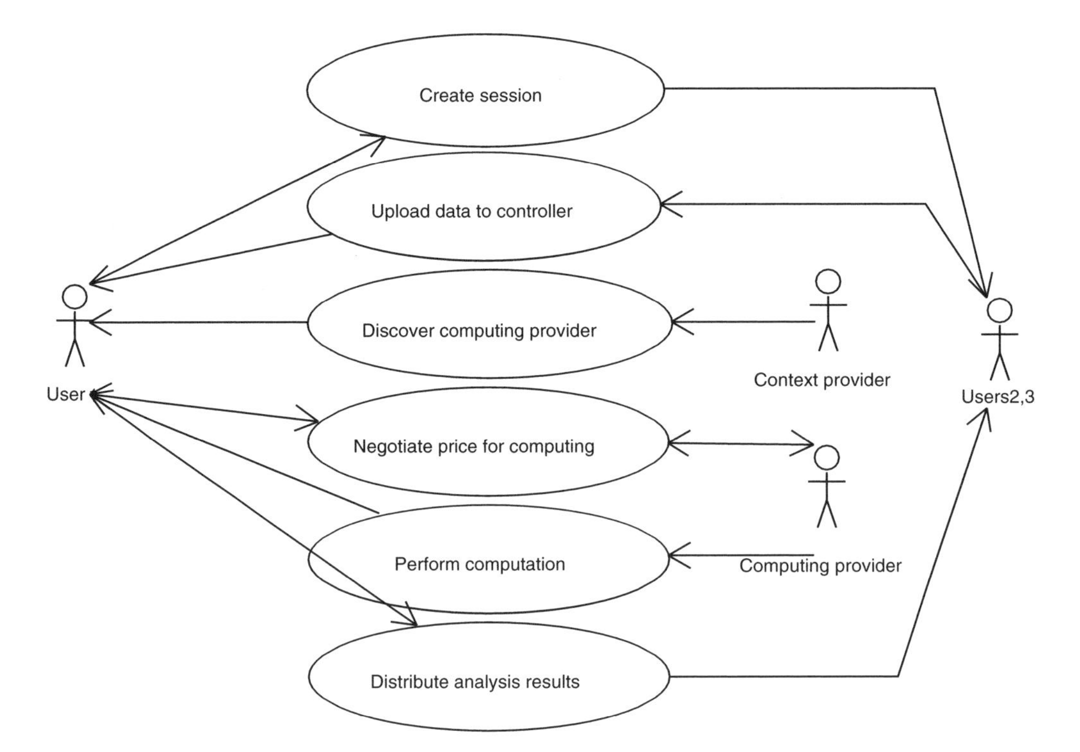

Figure 10.1 Use case view to distributed networking example

Another use case of potential interest is setting up the infrastructure for the previous use case. Such a use case would be useful for employer of users 1, 2, and 3. Nevertheless, We shall not consider it here.

10.4.2 Static View

Let us move on to describing the static view of the service model next. Figure 10.2 shows a subset of the constituents of the overall data analysis service.

Description of the entities in Figure 10.2:

- MixedP2PService: described previously
- P2PConnectivity: described previously
- WideAreaConnectivity: described previously
- DataAnalysisService: top-level view to data analysis service
- GroupSession: group session management
- DataUpload: upload of data from contributor to controller
- ResultDistribution: download of analysis results from controller to contributors
- Computation: performing of computation task.

An example of utilisation of Figure 10.2 is the observation that certain tasks (group session set-up, data uploading, results distribution) can be carried out without wide-area connectivity, whereas computation task does not require peer-to-peer connectivity. This kind of analysis could be carried out by the data analysis application automatically, so that data combining could be performed in *ad hoc* mode.

We shall next provide a model for the responsibility relations of the example. This is illustrated in Figure 10.3.

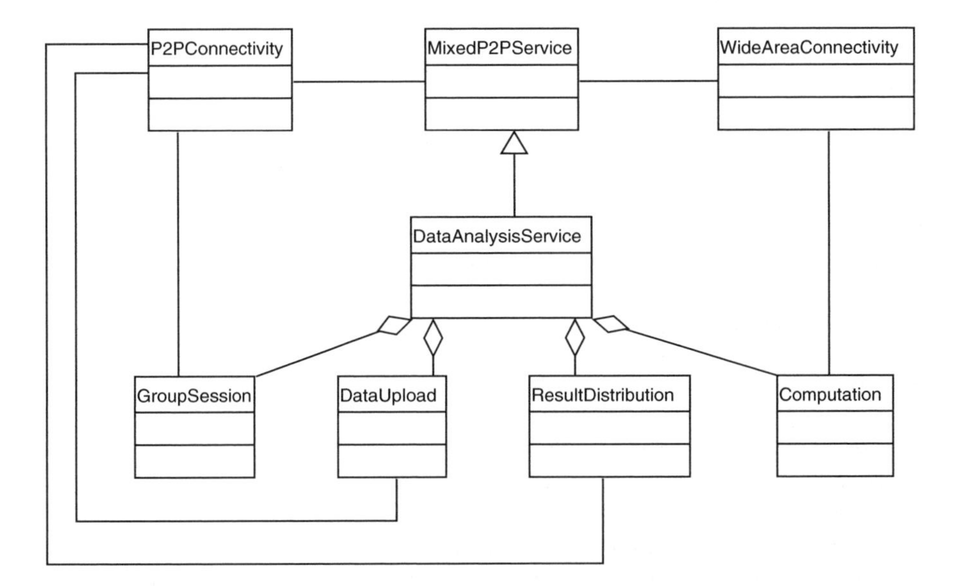

Figure 10.2 A subset of top-level service model for distributed service example

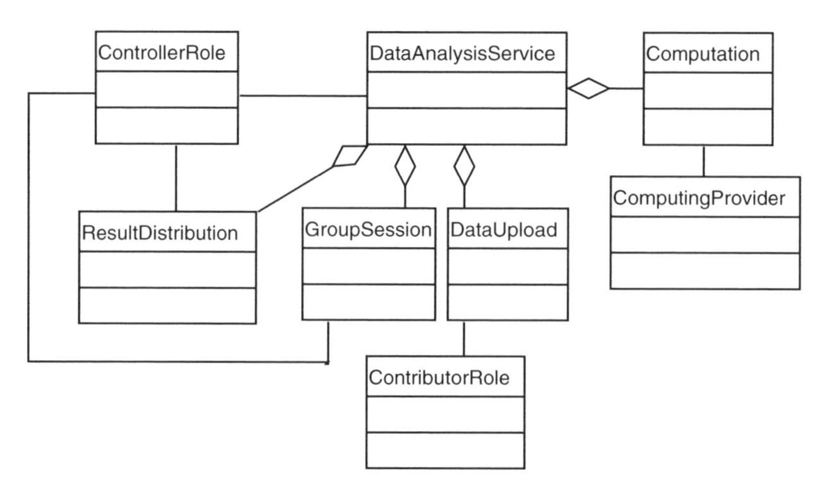

Figure 10.3 A model for responsibility relations in our example

Description of the entities in Figure 10.3:

- DataAnalysisService: described previously
- ResultDistribution, DataUpload, GroupSession, Computation: described previously
- ControllerRole: role of controller in data analysis service
- ContributorRole: role of contributor in data analysis service
- ComputingProvider: computing provider role in data analysis service.

We do not consider service quality related modelling for this example, since it has been part of the two previous examples.

10.4.3 Dynamic View

As an example of dynamic view, we illustrate the usage sequence of the service in Figure 10.4. Please note that we have omitted price negotiation with computing service provider for brevity. We have also omitted phases related to setting up of peer-to-peer communications.

Description of the entities and messages in Figure 10.4:

- User 1: described previously
- User 2 and 3: described previously
- Access provider: described previously
- Context provider: described previously
- ComputingProvider: described previously
- AdvGrp: advertise group communications
- JoinGrp: join group
- ActWAConn: activate wide-area communications
- AckWAConn: acknowledgment of wide-area communications activation
- SchProv: provider search

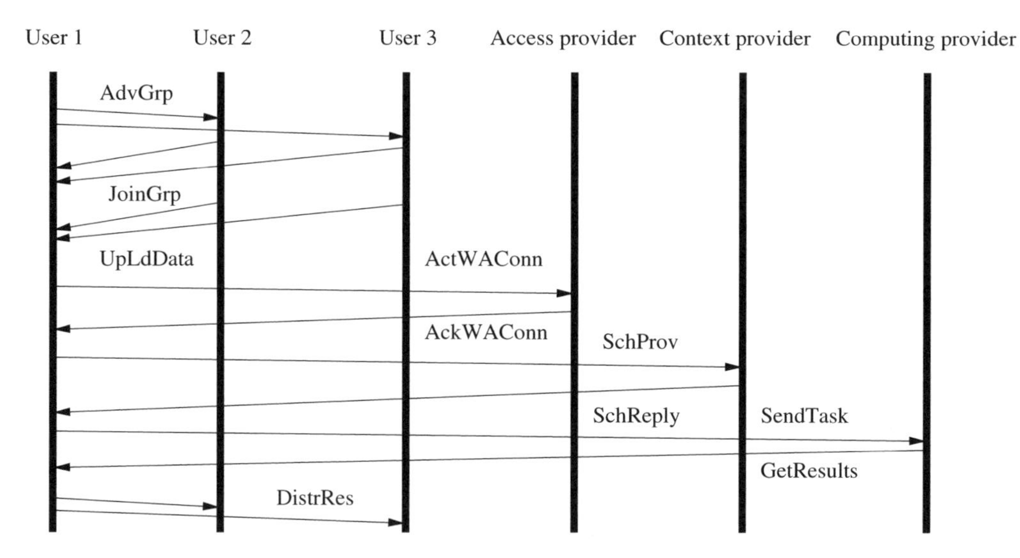

Figure 10.4 Sequence diagram for usage of the service

- SchReply: search reply
- SendTask: submission of computation task
- GetResults: transmission of computation task results
- DistrRes: distribution of the results.

Again, depending on the area of interest, a sequence diagram related to setting up the service could be useful for troubleshooting our example service by the help desk of the employer of the end-users.

10.5 Link to Service Management

We shall next consider service management aspects of our example. Our example is different from the previous ones because of the presence of non-managed components, and the fact that the overall responsibility for the service rests within an end-user. These facts are also reflected in service management, as we shall see.

10.5.1 Service Configuration

All told, service configuration for data analysis includes a number of different tasks. Let us study them according to the primary responsible stakeholder class. Please note that only user-related tasks have direct link to our example service, whereas others have an enabling role and may have been carried out at various points in time.
Service configuration tasks related to users:

1. Make sure that peer-to-peer communications capability exists for all group communications participants.

2. Ensure that group communications are supported by all participants.
3. Ensure that contributor members are able to export data and metadata in correct format.
4. Verify correct functioning of controller client.
5. Verify correct functioning of contributor clients.
6. Configure wide-area network configuration.
7. Ensure that context access functionality is operational.
8. Ensure that price negotiation functionality is operational.
9. Ensure uploading and downloading capability using interoperable data formats.

Service configuration tasks related to context provider:

1. Provision user 1 to customer database.
2. Enable repository access for customers.
3. Add computing provider to repository.

Service configuration tasks related to computing service provider:

1. Enable price negotiation functionality.
2. Enable computing facility.
3. Enable computation task metadata parsing functionality.

Service configuration tasks related to access provider:

1. Enable communications infrastructure.
2. Define QoS profiles per APN.
3. Define end-user classes.
4. Provision user 1 to customer database.

10.5.2 Service Assurance

As with service configuration, a number of different tasks can be read as belonging to service assurance. Most of them in our example relate to making sure that group-oriented communications operate the way that example service requires. Provider type stakeholders perform tasks in order to verify that the aspects of their externally provided services operate the way that they have been promised to.

In our example, the stakeholder responsible for end-to-end operation of the service is user 1. The most direct service assurance related task is observing the behaviour of the application during use. Various tools such as active measurements can be employed for probing technical operation of connectivity services.

10.5.3 Service Portfolio Management

Since our example consists of a single service only, end-user service portfolio management as such is not relevant for user 1. Ensuring that peer-to-peer connectivity and wide-area communication services most efficiently support the task at hand can be viewed to be a form of lower-level service portfolio management, supporting the use of the top-level service. Similar considerations relate to the version of the application used by controller and contributors for data analysis on one hand, and for group communications on the other.

Context provider, on the other hand, needs to monitor popularity of services and make sure that repository can meet the needs of the users.

10.5.4 Resource Development

For the user stakeholders, resource development activities relate to communications end-points. Context provider needs to ensure that resource capacity available is balanced with the demand. Computing service provider needs to consider the amount of CPU power available in view of the volume of computation tasks.

10.5.5 Product Management

Since user 1 is not selling data analysis service to users 2 and 3, business aspects of product management are not relevant here. Product management can be viewed to be still applicable in the sense that services are provided as a package associated with terms of use.

Context provider needs to package repository access into the form or product, and computing provider needs to do the same for computing service. Access provider makes a product out of wide-area connectivity, in our case also including access to context provider's services. Context provider's services may also be used as a separate product by other parties.

10.5.6 Use of Policies

As described earlier, individual users can employ preferences for peer-to-peer communications for the purpose of automating related activities. We shall not discuss how provider stakeholders could use policies for data analysis service, since examples of this have been provided previously.

10.6 Summary

Our example showed how service modelling can be used in the context of mixed peer-to-peer/managed services. Service modelling can be used for ensuring that all functionalities are in place. In contrast to previous examples, service modelling in this case would most likely be used by other parties than the service provider (user 1). For example, the IT department of the employer of end-users could make use of it to ensure the presence of all the necessary functionalities.

This kind of future scenario provides lots of interesting potential uses for service modelling. Service models, in one form or another, can be expected to form a basis for automating of interactions between stakeholders.

10.7 Highlights

Ten things to remember from this Chapter:

- Distributed networking example involves both *ad hoc* connectivity and wide-area connectivity.
- Same scheme would also support remote peer-to-peer participation.
- One of the end-users (controller) is responsible for the overall service.
- Other end-users (contributors) use the service provided by controller.
- Contributors make use of controller's subscriptions.
- Service makes use of group communications.
- Utility computing is used for remote processing of data.
- Metadata is used for exchanging data between contributor and controller, and for submitting utility computation task.
- Utility computation task is described in a generic form.
- Results of the computation are distributed to contributors by controller.

Part IV
Summary

Scope of Part Four

In this part, we shall summarise the essential learnings from the first Three Parts and outline issues that are relevant for the future.

11

Summary

We have reached the final chapter of the book and we shall first take a moment to review the most important topics covered. The topic area, service modelling, being relatively novel, large, and encompassing technologies under development, Summary is intended to put the topics covered in perspective.

We have discussed state of the art in providing packet-based services, and identified issues of importance which have relevance to the topic of this book. Not only are systems and services getting ever more complex, but also the way business is done is undergoing a change. Partly a result of tougher competition and partly facilitated by technological advances, enterprises can form value nets more easily than before, targeting increase in operational effectiveness and sharpening of business focus. Making true of this promise, sets new level of requirements for infrastructure, underlining the importance of open standardisation and collaboration. The importance of cooperative concept work has been demonstrated by the success of Global System for Mobile communications (GSM) standard, and can be expected to still increase.

Getting to the actual topic area of this book, the importance of management software configurations is the next step in the shifting of focus from hardware to software. A promising way to manage complexity is that representation of capabilities of servers and systems as services, bringing structure to configuration management. Services can make use of other services, link to resources, and support functionalities visible to end-users. This kind of approach facilitates reuse. Services can be viewed to have a life cycle associated with them, and can be viewed as being related to service management processes. These considerations led us to two important viewpoints on service modelling: service topology on one hand and process modelling on the other. We reviewed modelling methods for static information and processes, and noted that some of the software development practices carry over to service modelling and service management even if there would be no actual software being developed.

A number of industry initiatives were reviewed in view of their relevance to service modelling. Standardisation bodies such as Third Generation Partnership Project 3GPP and Internet Engineering Task Force (IETF) provide architectures and protocols for connectivity supporting service provisioning and use of services. 3GPP and Open Mobile Alliance (OMA) work on standardisation of the platforms on which services are operated, exemplified by IP Multimedia Subsystem (IMS) and OMA Service Provider Environment (OSPE). Other fora work on a more abstract level. Object Management Group (OMG) has specified the UML (Unified Modelling Language) language and methodology modelling, and Institute for Electrical and Electronics Engineers (IEEE) has described architectural practices to be followed in modelling. TeleManagement Forum has done work in the areas of process and information modelling, including service modelling. Different fora are interacting with future-oriented research project such as EU FP6 projects and Wireless World Research Forum (WWRF).

We next summarised key requirements for service modelling in a generic scope. We used a number of viewpoints to organise concerns and requirements, with stakeholder-specific viewpoints having an important role.

A management operations framework was described providing a setting for interactions between stakeholders on one hand and service management related operations within stakeholders on the other. Service life cycle tasks are governed by the requirements of business transactions between stakeholders. Service model facilitates effective operations for a stakeholder, and is also an important enabler for inter-stakeholder exchanges. Different service life cycle operations can be linked to particular roles, or sets of tasks, providing a way to map service life cycle to management framework. A part of the service model, service framework, decomposes the technical definition of a service into a number of entities and their inter-relations. The use of service framework in connection with end-user services was illustrated with examples.

We reviewed a number of modelling patterns illustrating issues of the use of service modelling, followed by a discussion about steps required to construct a complete service model for a particular operations environment. Service modelling patterns were illustrated and complemented by three examples, relating to DiffServ networking, mobile networks, and distributed systems.

One of the topics emerging from among the topics covered is the importance of service level management in moving toward architecture based on the representation of capabilities as services. When services are aggregated, service level definitions pertaining to different parts of the hierarchy are of great importance. In distributed systems, modelling of service level for connectivity between service functionalities brings a further viewpoint to the picture.

11.1 Issues for the Future

Multiplicity of technologies in the market seems to be again on the increase, after a period of slower launches that lasted for a few years. For example, utility computing and grid computing have attracted attention. Nevertheless, it is most likely that the greatest, or at least the most useful, of innovations in the near future result from putting the known technologies and theoretical paradigms together in a form that brings added value to the end-user, can be explained to end-users in practical terms, and does not increase apparent complexity of the service usage.

Service modelling in making more advanced capabilities available, without making life overly difficult – either for the end-user or for the personnel responsible for creating and operating the capabilities. Service modelling is already used by connectivity and service providers for facilitating introduction of new services and managing them together. Service modelling may be expected to play a role for easing the life of the end-users, but this act most likely takes place behind the curtain, so to speak. Context-sensitive service platforms of the future can utilise service models to reduce apparent complexity of end-user services. This can be compared to the fact that early automobiles had a control for ignition adjustment next to the steering wheel, whereas corresponding functionality is automated in modern vehicles. User is presented with more useful information instead, for example, a reminder of impending regular maintenance.

It is interesting to see what the future brings with it.

Part V
Appendices

3GPP Bearer Concepts

In the following text, we shall review 3GPP (Third Generation Partnership Project) bearer concepts and their relation to service quality provisioning. There are minute variations of the details of how exactly this takes place, but we shall describe here a streamlined version that roughly corresponds to 3GPP Release 5. The description is based on 3GPP as described in (3GPP TS 23.107, 2004) and (3GPP TS 23.207, 2004).

The 3GPP architecture is based on end-to-end bearers between the end-user and the service. The architecture has been designed to support different kinds of services having specific service quality requirements and characteristics, and supports the simultaneous use of multiple services.

In principle, the high-level 3GPP architecture supports end-to-end negotiation involving signalling external to the mobile network also. The high-level reference picture for 3GPP is shown in Figure A.1. In practice, bearer related signalling takes place only within the network and external service quality is handled with Service Level Agreements (SLAs).

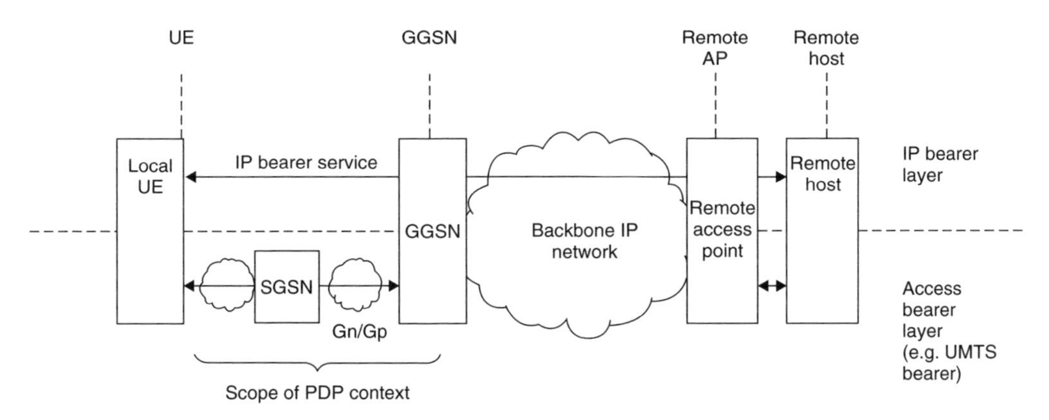

Figure A.1 High-level end-to-end reference model for 3GPP QoS architecture. From (3GPP TS 23.107, 2004)

Service quality within a cellular domain is based on negotiated bearers between end-user equipment and the network. In the current embodiment of the standard, the end-user equipment is responsible for activating the bearer. The range of allowable bearer parameters is provisioned by the network provider. In principle, provisioning could be performed on user and service granularity, but in practical implementations the range is provisioned for a class of services and for a group of end-users.

For accessing a particular service, a new bearer can be activated, or a service can be used by modifying an existing one. It is possible to modify the properties of the bearer for the latter purpose. It is useful to note that using an existing bearer may involve multiplexing of multiple non-related service flows onto the same bearer.

In order to understand 3GPP service quality support adequately, it is useful to have a basic understanding of the 3GPP Quality of Service (QoS) architecture. Consequently, we shall review the central architectural concepts prior to discussing bearers and service provisioning.

Architecture

The top-level conceptual illustration of the 3GPP QoS architecture is shown in Figure A.2. It consists of a hierarchy of bearers between functional blocks of the architecture. We shall summarize functional blocks and bearers in the following text, and proceed to describe how different bearers relate to each other.

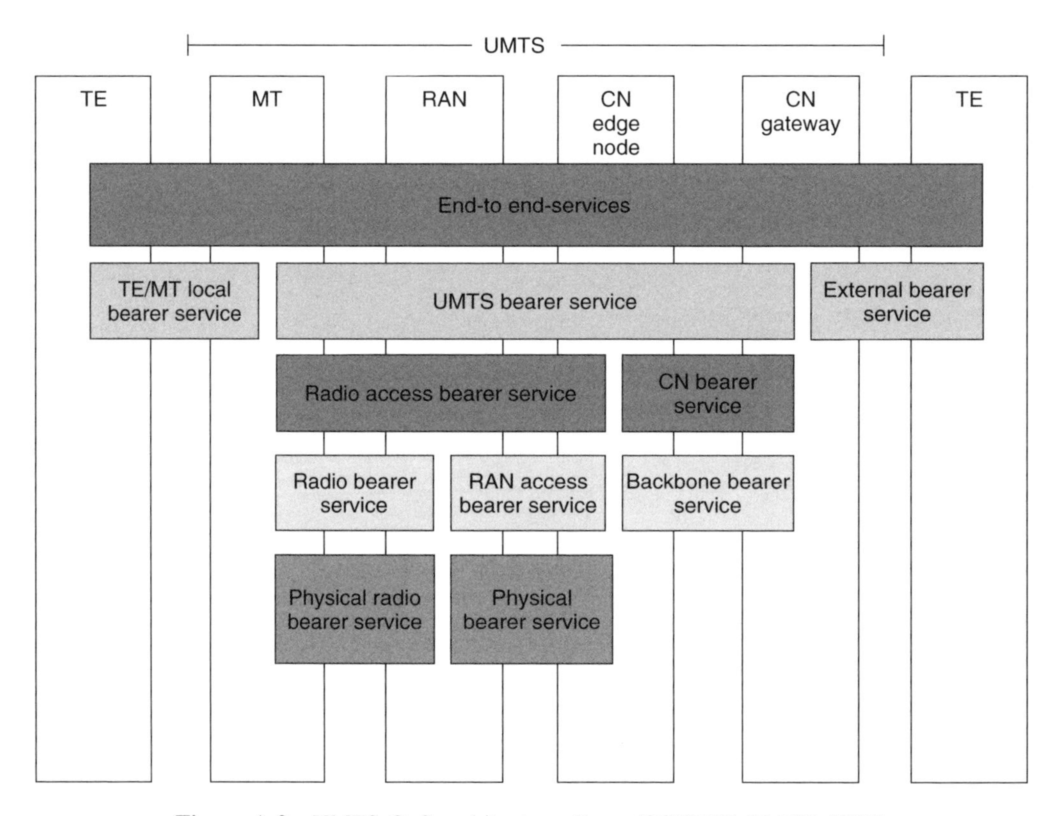

Figure A.2 UMTS QoS architecture. From (3GPP TS 23.207, 2004)

The functional blocks in the figure are as follows:

- TE (left-hand side): Terminal equipment. Can be, for example, laptop computer, Internet tablet, or cellular phone itself
- MT: Mobile terminal. Functionality which terminates the Universal Mobile Telephony System (UMTS) bearer service. MT handles one end of the UMTS bearer negotiation
- RAN: Radio Access Network. This functionality controls access to radio interface resources and handles most of the mobility support such as handovers between base stations (called *Node B*s for Wideband Code Division Multiple Access (WCDMA)
- CN Edge Node: Terminates Radio Access Bearer Service. In practice means Service Gateway GPRS Support Node (SGSN) in the current architecture
- Core Network (CN) Gateway Node: Endpoint of UMTS bearer service in the network. The master node for UMTS bearer negotiation in the network. In practice, GPRS Gateway Support Node (GGSN) fulfils this function
- TE (right-hand side): Another terminal or service.

The topmost horizontal entity in the figure is end-to-end service, and describes the conceptual bearer between end-user and the other end of the communication. It is supported by three lower-level bearers, namely, 'TE/MT Local Bearer Service', 'UMTS Bearer Service', and 'External Bearer Service'. These relationships mean that the lower level bearer services are instantiated in such a way as to produce the desired properties for the higher-level bearer service using it. In the same way, UMTS bearer service is supported by 'Radio Access Bearer Service' and 'CN Bearer Service', which – in turn – can be analysed in terms of lower-layer bearers. Eventually, the bearers map to resources which provide adequate service quality support.

We shall not need to go through all of the bearers; for the present purpose it is sufficient to understand the principle. We shall, however, study the second-highest level since there are certain important aspects related to it.

The UMTS bearer provides the mobile network support for service quality. This is the bearer which is negotiated in practice. To achieve adequate end-to-end service quality, the rest of the bearers need to match the end-to-end performance targets, too. In particular, this is relevant for the external bearer linking the external service (or other terminal) and the mobile network. In practical implementations, multiple service quality support classes can be implemented with DiffServ towards non-cellular networks and with General Packet Radio System (GPRS) roaming exchange mechanisms towards cellular networks. The UMTS bearer resulting from negotiation is mapped to a suitable external service quality support class. We shall not discuss the local TE/MT bearer here, since it is typically outside of operator control.

The 3GPP architecture document provides an overview of the logical functionalities involved in UMTS bearer negotiation in the form of two pictures which are shown in the following text.

Figure A.3 shows the 'control layer', or functionalities which are involved in setting up the UMTS bearer. Each of the architectural functional blocks involved in UMTS bearer is associated with admission control and bearer service managers. Conceptually, multiple levels of bearers are associated with a manager of functionalities, so that the UMTS bearer manager is connected to the RAB manager and core network bearer manager, for

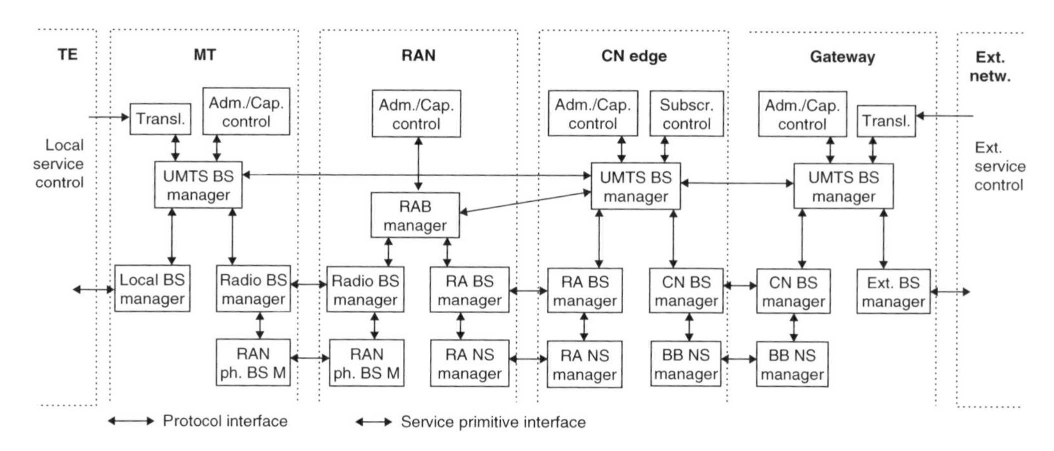

Figure A.3 Control layer functionalities for UMTS bearer. From (3GPP TS 23.207, 2004)

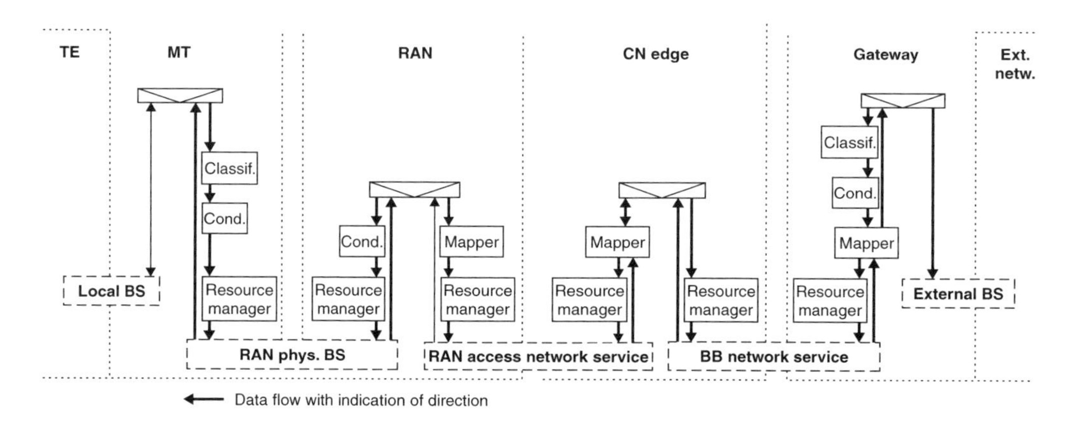

Figure A.4 User layer functionalities for UMTS bearer. From (3GPP TS 23.207, 2004)

example. This figure illustrates nicely the central role played by GGSN in UMTS bearer negotiation.

Figure A.4, in turn, illustrates the 'user layer' functionalities belonging to the UMTS bearer, corresponding to treatment packets that are subjected along the UMTS bearer data path. The central idea here is providing of service quality support between the functional blocks. Within each block, the following operations can be applied to packets making up the user layer traffic:

- Classification: detect suitable service quality support for the packet in question
- Conditioning: application of operations such as traffic shaping and/or buffering for controlling token bucket parameters
- Mapping: mapping packet to adequate traffic aggregate corresponding to the classification.

Resource manager takes care of low-level access to transport resources. Note that in the preceding text we have spoken of per-packet operations in the spirit of Internet routing. In practice, one needs to take into account flow orientation.

As discussed in Chapter 1, one can differentiate between static and dynamic provisioning for services. 3GPP architecture supports both, providing the ability to dynamically link bearer properties to service usage session as shown in Figure 1.4.

Bearer negotiation

Bearer negotiation takes place between the terminal and network, and is associated with an Access Point Name (APN). An APN is essentially a Service Access Point (SAP). Bearer negotiation may result from bearer activation or bearer modification.

Bearer activation procedures are somewhat different for static and dynamic provisioning modes described in Chapter 1. We shall discuss them next.

In static mode, bearer activation for a specific APN is requested by the terminal. The terminal may provide QoS attributes as part of the bearer activation or modification request. If attributes are not provided, network will fill them in. Network may downgrade requested QoS attribute values, but cannot upgrade them.

In dynamic provisioned mode, some of the QoS attributes are determined by a service session, for example, media stream belonging to a Session Initiation Protocol (SIP) multimedia session. In this example case, IP Multimedia Subsystem (IMS) provides a special authorisation token to be supplied as part of bearer activation request. GGSN uses the token to check that requested attributes indeed do correspond to session parameters.

The QoS attributes that can be associated with a bearer are shown in Table A.1. Some of the attributes are relevant only for some of the traffic classes, as indicated in the table. For example, guarantee-type attributes are only relevant for real-time traffic classes (conversational and streaming) for the reason that they need constant token rate. Interactive traffic class is suitable for browsing, and does not provide bandwidth guarantees in the standard implementation. Traffic Handling Priority (THP) is provided as a means of prioritizing Service Data Units (SDUs) within the interactive traffic class.

Table A.1 3GPP bearer QoS attributes and their relevance to the four 3GPP traffic classes (conversational, streaming, interactive, and background)

Attribute	Conv	Str	Int	Bg
Maximum bit rate	X	X	X	X
Delivery order	X	X	X	X
Maximum SDU size	X	X	X	X
SDU format information	X	X		
SDU error ratio	X	X	X	X
Residual bit error ratio	X	X	X	X
Delivery of erroneous SDUs	X	X	X	X
Transfer delay	X	X		
Guaranteed bit rate	X	X		
Traffic handling priority			X	
Allocation/retention priority	X	X	X	X
Source statistics descriptor	X	X		
Signalling indication			X	

Service provisioning

Static and dynamic provisioning modes were described in Chapter 1. They correspond to primary and secondary APNs in GGSN. We shall concentrate on primary APN provisioning in the following text.

An APN represents a provisioning point for network operator, allowing for application of service class specific policies mapped to an APN. One of the most important policies is service quality control. Maximum service quality level is provisioned for end-users per APN in Home Location Register (HLR). In the context of bearer negotiation, in addition to APN-specific 'maximum service quality', resource availability is taken into account. In addition to service quality, an APN can be associated with other parameters as well, including security and IP tunnelling related ones.

For network operator, an APN is a central means of affecting service quality and security support for services. From the viewpoint of IP-based end-user services, it could be viewed as being associated with logical resource processes and having the nature of an enabler. It is useful to keep in mind that access to Internet could also be viewed as a product-facing service.

B

DiffServ SLA Concepts

In this Appendix, we shall describe some of the DiffServ Service Level Agreement (SLA) related concepts used in the book in more detail. We shall start with a short DiffServ primer, proceed to describe DiffServ SLA concepts, and end with Per-domain Behaviours (PDBs). DiffServ SLA terminology description is based on (Grossman, 2002) and PDB text on (Nichols and Carpenter, 2001).

DiffServ primer

DiffServ is a method for providing statistical service quality support within a network domain. It does not make assumptions about service quality support along the entire end-to-end delivery chain and does not require instantiation of service quality support. Consequently, no session-specific resource reservation is made in the network domain, and no dedicated functionality in the communications endpoint is required.

In DiffServ framework, forwarding nodes (routers) are divided into edge routers providing special treatment at network ingress and egress on one hand, and core routers performing only forwarding on the other hand. The rationale for the set-up is provided by the fact that typically core routers aggregate traffic from multiple access routers (possibly in multiple stages), whereby the relatively large number of access routers facilitates more complex operations at the domain edge as compared to core routers. Figure B.1 illustrates the conceptual construction of a DiffServ domain.

The need for service quality arises from the need to provide service quality for multiple customers simultaneously while striving to use network resources effectively. The simplest but costliest solution would be to build in so much capacity that forwarding resources are never saturated, no matter how much traffic enters the network domain. The DiffServ approach allows for the use of smaller installed bases by leveraging the fact that different traffic flows have different requirements regarding service quality. Figure B.2 provides an example of this.

In the example of Figure B.2, traffic from user 1 to user 2 traverses routers A, F, G, and C. Traffic from user 3 to user 4 traverses routers B, F, G, and D. Thus, link F-G represents shared capacity between these two traffic aggregates and egress scheduler of router F towards link F-G needs distribute link capacity between these two aggregates in our example. If both traffic aggregates contain flows which are more urgent than other

Service Modelling: Principles and Applications
Vilho Räisänen

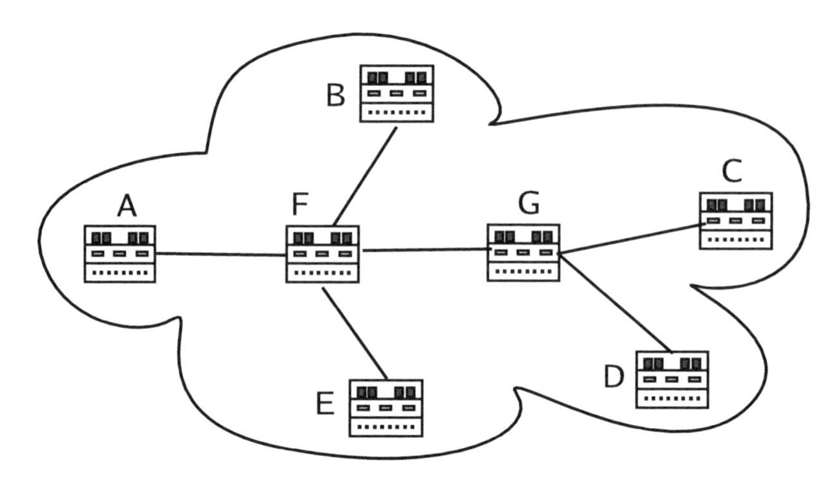

Figure B.1 An illustration of the construction of a DiffServ domain. A–E are edge routers, whereas F and G are core routers

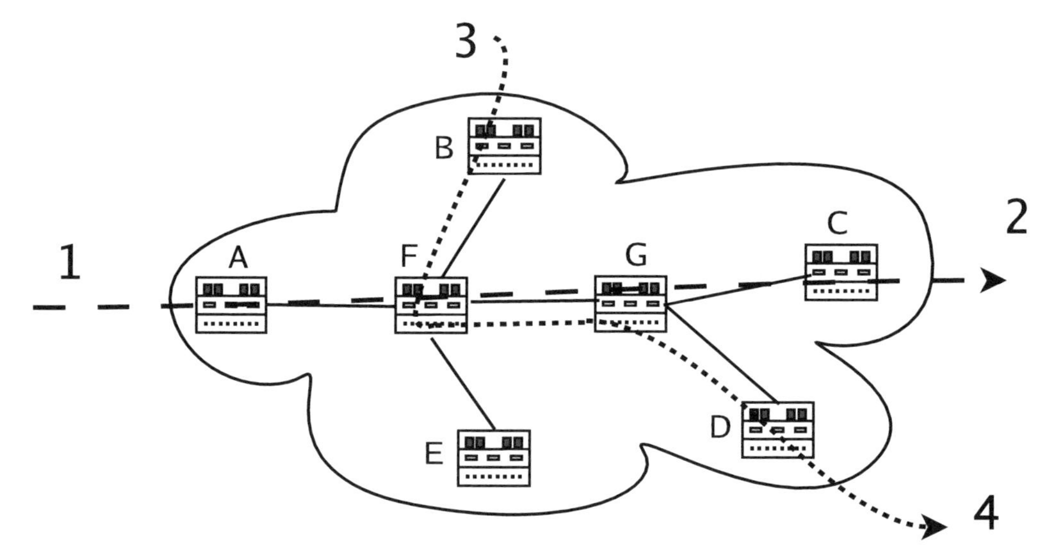

Figure B.2 Example of sharing capacity. User 1 sends traffic to user 2 and user 3 to user 4

ones, the former can be scheduled with higher priority than the latter. This is the basic idea of DiffServ as with other differentiated forwarding frameworks.

The DiffServ framework is based on classifying and conditioning traffic at the edge of the network domain and providing differentiated treatment within the domain based on the classification. Let us next study different phases experienced by an Internet Protocol packet as it traverses the DiffServ network domain. In the spirit of Internet, DiffServ is packet-oriented rather than session-oriented.

A rough overview of the phases looks like this:

- Classification
- Per-hop Behaviour (PHB) allocation

- Ingress conditioning
- Forwarding
- Egress conditioning.

We shall discuss different phases below.

Classification is the basis for allocating service quality support for the packet in the DiffServ domain. Classification can be based on multiple criteria, such as the type of the service in question or subscriber. Since there is no flow-specific negotiation involved, classification is based on the information within the header of the IP packet. Information which can be used as a basis of classification includes

- Source IP address
- Destination IP address
- DiffServ Code Point (DSCP) field
- Protocol number
- Flow label (IPv6).

Source IP address can be used as a basis for identifying the sender, possibly also for identifying service. Destination IP address can be used for identifying the recipient of the packet. Protocol number information can provide hints about Layer 4+ (L4+) protocols and hence about services involved. Tunnelling of packets, as well as encryption, pose challenges to applying this method. Flow label is only supported in IP version 6 (IPv6), and provides a means of identifying individual flows within a DiffServ domain.

For sender or recipient identification, subscription specific SLA can be used as a basis for determining classification. Operator may have service specific policies, in enforcing which, information about service can be used. To prevent unauthorised use of resources, typically service based policies need to be complemented with relevant subscription identification. In some cases such as Voice over IP (VoIP), session-specific gating, analogous to one used in IP Multimedia Subsystem (IMS), may be employed but we shall not discuss it here further.

Based on classification, a packet is allocated a PHB. DiffServ standard framework describes the following PHBs:

- Expedited Forwarding (EF) PHB
- Assured Forwarding (AF) PHB group
- Best-effort (BE) PHB.

The chosen PHB is reflected in DSCP marking in the packets. Let us review these next as they determine the service quality support within the DiffServ domain. More detailed summary can be found for example in (Räisänen, 2003a).

Expedited Forwarding is meant to be used as a building block for low-loss, low-latency, guaranteed bandwidth end-to-end delivery service. It is suitable for services with strict inherent service quality requirements such as VoIP media streams.

Assured Forwarding PHB group comes in no less than 12 variants, which are organised into four AF classes, each consisting of three drop precedence levels. An AF class is meant to be associated with an allocation of traffic forwarding resources such as buffer space and egress interface link capacity in individual routers. The four classes provided do not have any inherent meaning as such. The drop precedence level indicates which packets within an AF class are discarded first, should shortage of buffer space occur. Roughly speaking, forwarding class can be associated with scheduler parameters and precedence level with buffer management parameters.

Best-effort PHB does not have any service quality support behaviour defined with it, and can be used for transporting packets which do not belong to other PHBs.

Traffic conditioning is an operation which can be used in both domain ingress and domain egress. In the network ingress, the purpose is to alleviate non-desirable effects such as periods of heavy packet loss and/or increased latency by smoothing traffic variations. Traffic shaping with buffering and packet dropping can be used to this end. When used in network egress, the purpose of traffic conditioning is to make traffic towards other domains conformant with relevant Traffic Conditioning Agreements (TCAs).

All of the operations above are performed at network egress nodes. Traffic forwarding in network core routers is based on normal IP routing supplemented by per-node operations relating to PHB as detected by DSCP marking in the packet header. The DiffServ framework does not define the actual implementation of forwarding within the core routers, but rather describes the effect that should result from the use of selected mechanisms. Often, priority scheduler is used for EF PHB and for example in Weighted Fair Queuing (WFQ) for AF PHB group. As described above, buffer management algorithm is needed for drop precedence management.

The standard DiffServ framework is based on static provisioning paradigm, where classification, conditioning, and forwarding configurations are made off-line. The research community has developed dynamic provisioning schemes, usually called *bandwidth brokers*, to complement static DiffServ scheme with instantiated service quality support schemes.

The transport provider may employ traffic engineering methods to provide added control in a DiffServ domain. For example, routing of packets can be controlled with MPLS Label-switched Paths (LSPs). Traffic can be mapped to an LSP based on the same classification criteria as DiffServ, and can hence support PHB-specific LSPs also. Different LSPs between same endpoints may be routed differently, and be used for example for resilience purposes, in addition to PHB-specific routing.

More information about bandwidth brokers and traffic engineering can be found for example in (Räisänen, 2003a).

Let us next move on to describing DiffServ SLA concepts.

DiffServ SLA

The DiffServ SLA is a set of concepts describing how SLAs can be implemented and used in a DiffServ domain. The DiffServ SLA framework consists of the following concepts:

- Service Level Agreement (SLA)
- Service Level Specification (SLS)
- Traffic Conditioning Agreement (TCA)
- Traffic Conditioning Specification (TCS).

Let us take a look at these next.

SLA is a service contract between DiffServ transport provider and a customer, specifying the forwarding service the customer is entitled to.

Service Level Specification is a set of technical parameters providing detailed information about the forwarding service within a DiffServ domain. A SLS may contain PDB definitions relating to traffic aggregates.

Traffic Conditioning Agreement describes classifier rules as well as metering, marking, discarding, and shaping rules relevant to traffic profiles.

Traffic Conditioning Specification contains technical parameters relevant to traffic conditioning.

A SLA typically includes different parts relating to applicability, supervision, and reporting, which we shall not discuss here. The reader is referred for examples to Räisänen (2003a) for more information.

Per-domain behaviours

We have seen above that DiffServ is based on statitical service quality support determined by the PHB allocated for the traffic in question. The contribution of the DiffServ domain to end-to-end service quality is affected by the following factors:

- Domain-internal parameters: configurations of classification, conditioning, PHB allocation, and traffic forwarding
- Characteristics of the customer's traffic
- Other traffic sharing the traffic aggregate.

The reason why domain-internal parameters affect the per-domain service quality support should be clear after the previous description of DiffServ fundamentals.

Characteristics of customer's traffic entering the DiffServ domain determine the conditioning applied to it at network ingress, potentially affecting the per-domain performance.

Other customers' traffic sharing aggregates affects the forwarding in individual core network routers.

Traffic conditioning agreement clarifies the effect of the second of these factors. The customer does not need to know the details pertaining to the first and the third one, but needs to know what is the net result on its own traffic. DiffServ Per-domain Behaviours (PDBs) have been defined to meet this need.

A PDB describes the expected treatment that an identifiable set of packets will experience, edge to edge, in a DiffServ domain. To be useful, a PDB must describe expected treatment in the form of measurable entities. Typical characteristics relevant to PDBs include packet loss rate and latency. Characteristics are normally described using statistical statements such as 'for 99% of packets, ...'. The PDBs may be defined network-wide, or between Points of Presence (PoP).

Since PDB is described essentially in terms of measurements, the methodology employed in measuring characteristics is of great importance. Issues to consider include: measurement method (active or passive), the way sampling is performed, the length of an individual measurement. The IP Performance Measurement (IPPM) working group of Internet Engineering Task Force (IETF) has developed a measurement framework as well as a suite of measurement methods which are most likely useful in this context. Poisson sampling may provide most general results for a heterogeneous traffic mix, whereas periodic measurement may yield most information for specific applications such as VoIP.

Bibliography

3GPP TS 23.107, *Quality of Service (QoS) Concept and Architecture*, version 5.13, December 2004.

3GPP TS 23.207, *End-to-end Quality of Service (QoS) Concept and Architecture*, version 5.9, December 2004.

A conversation with Roger Sessions and Terry Coatta, *ACM Queue* **3**, issue 7, p. 16ff., 2005.

Aftelak A., Häyrynen A., Klemettinen M., and Steglich S., *MobiLife: applications and services for the user-centric wireless world*, IST Mobile and Wireless Communications Summit 2004, Lyon, France.

Ahmavaara K., Haverinen H., and Pichna R., Interworking architecture between 3GPP and WLAN systems, *IEEE Communications Magazine* **41**, p. 74 ff., 2003.

Ambient: please see project home page at `http://www.ambient-networks.org`, February 2006.

Armitage G., *Quality of Service in IP networks*, MacMillan Technical Publishing, Indianapolis, USA, 2000.

Berners-Lee T., Hendler J., and Lassila O., *The Semantic Web*, Scientific American, 2001.

Black D., Blake S., Carlson M., Davies E., Wang Z., and Weiss W., *An Architecture for Differentiated Services*, RFC 2475, IETF, December 1998.

Bouch A., Sasse M., DeMeer H., Of packets and people: a user-centered approach to Quality of Service, *Proceedings of the IWQoS'00*, IEEE, Pittsburgh, USA, June 2000.

Braden R., Clark D., and Shenker S., *Integrated Services in the Internet Architecture: An Overview*, RFC 1633, IETF, June 1994.

Brereton P., The software customer/supplier relationship, *Communications of the ACM* **47**, p. 77 ff., February 2004.

cdma2000 Evaluation Methodology, revision 0, version 1.0, C.R1002-0, 3GPP2, December 2004.

Churchill E., Girgensohn A., Nelson L., and Lee A., Blending digital and physical spaces for ubiquitous community participation, *Communications of the ACM* **47**, p. 39 ff., February 2004.

Common Object Request Broker Architecture: Core Specification, OMG, March 2004.

Cortese G., Fiutem R., Cremonese P., D'Antonio S., Esposito M., Romano S.P., and Diaconescu A., Cadenus: creation and deployment of end-user services in premium IP networks, *IEEE Communications Magazine* **41**, p. 54 ff., 2003.

DAML-S: Semantic Markup for Web Services, version 0.9, the DAML services coalition, DARPA, May 2003.

Davies N., Fensel D., and Richardson M., The future of web services, *BT Technology Journal* **22**, p. 118 ff., January 2004.

de Marca J., Tafazolli R., and Uusitalo M., WWRF visions and research challenges for future wireless world, a series of articles within *IEEE Communications Magazine* **42**, 2004.

E2R: please see project web site at `http://e2r.motlabs.com`, February 2006.

ETSI, *End-to-end Quality of Service in TIPHON Systems, Part 2: Definition of Speech Quality of Service (QoS) Classes*, TS/TIPHON 101329-2, 2000.

Enhanced Telecom Operations Map (eTOM), version 4.5, GB 921, TMF, December 2004.

eTOM Application Note V: an Interim View of an Interpreter's Guide for eTOM and ITIL Practitioners, GB 921V, TMF, February 2005.

Ferguson D., Sairamesh J., and Feldman S., Open frameworks for information cities, *Communications of the ACM* **47**, p. 45 ff., February 2004.

Foster I., Service-oriented science, *Science* **308**, p. 814 ff., 2005.

Gamma E., Helm R., Johnson R., and Vlissides J., *Design Patterns — Elements of Reusable Object-Oriented Software*, Addison-Wesley, Indianapolis, USA, 2004.

Goldstein H., Who killed the virtual case file? *IEEE Spectrum*, September issue, **42**, p. 18 ff., 2005.

Grossman D., *New Terminology and Clarifications for DiffServ*, RFC 3260, IETF, April 2002.

Halonen T., Romero J., and Melero J., *GSM, GPRS, and EDGE Performance — Evolution Towards 3G/UMTS*, John Wiley & Sons, Chichester, England, 2003.

Handley M., Schulzrinne H., Schooler E., and Rosenberg J., *SIP: Session Initiation Protocol*, RFC 2543, IETF, March 1999.

Heckmann O., Rohmer F., and Schnitt J., *The token bucket allocation and reallocation problem*, `http://www.kom.e-technik.tu-darmstad.de/publications/abstracts/HRS01-1.html`, 2002.

Henderson-Sellers B., Understanding metamodelling, in *Proc. ER2003*, Chicago, USA, October 2003. Home page at `http://www.er.byu.edu/er2003/`.

Hill J., A management platform for commercial web services, *BT Technology Journal* **22**, January 2004.

Hollander A., Denna E., and Cherrington J., *Accounting Information Technology, and Business Solutions*, McGraw Hill, Singapore, 2000.

Hollingsworth D., *The Workflow Reference Model*, TC00-1003, Workflow Management Coalition, 1995.

IEEE Recommended Practice for Architectural Description of Software-Intensive Systems, IEEE standard 1471-2000, 2000.

Introductory Overview of ITIL, `http://www.itsmf.com`, itSMF, 2004.

ITU-T Recommendation Y.110, *Global Information Infrastructure Principles and Framework Architecture*, June 1998.

ITU-T Recommendation G.109, *Definition of Categories of Speech Transmission Quality*, September 1999.

ITU-T Recommendation G.1000, *Communications Quality of Service: A Framework and Definitions*, November 2001.

ITU-T Recommendation G.1010, *End-user Multimedia QoS Categories*, November 2001.

ITU-T Recommendation G.809, *Functional Architecture of Connectionless Layer Networks*, March 2003.

Jones S., Toward an acceptable definition of service, *IEEE Software* **22**, p. 87 ff., 2005.

Kelly F., Models for self-managed Internet, *Philosophical Transactions of the Royal Society* **A358**, p. 2335 ff., 2000.

Kilkki K., *Differentiated Services for the Internet*, MacMillan Technical Publishing, Indianapolis, 1999.

Klemm A., Lindemann C., and Lohmann M., Traffic modelling and characterization for UMTS networks, *Proceedings of the GLOBECOM'01*, IEEE, 2001.

Koivukoski U. and Räisänen V. (editors), *Managing Mobile Services – Technologies and Business Practices*, John Wiley & Sons, Chichester, England, 2005.

Koodli R. and Puuskari M., Supporting packet-based data QoS in next-generation cellular networks, *IEEE Communications Magazine* **39**, p. 180 ff., February 2001.

Laiho J. and Acker W. (editors), *WCDMA for UMTS*, 2nd edition, John Wiley & Sons, Chichester, England, September 2005.

Lakaniemi A., Rosti J., and Räisänen V., Subjective VoIP speech quality evaluation based on network measurements, *Proceedings of the ICC'01*, IEEE, Helsinki, Finland, 2001.

Lassila O. and Dixit S., Simple approach to automatic service substitution, in *Proc. AAAI Spring Symposium on Web Services*, Stanford, USA, 2004.

Leung K., Massey W., and Whitt W., Traffic models for wireless communications networks, *IEEE Journal on Selected Areas of Communications* **12**, p. 1353 ff., 1994.

Liberty: please see web site at `http://www.projectliberty.org`, February 2006.

Martin-Flatin J-P., Srivastava D., and Westerinen A., Iterative multi-tier management information modelling, *IEEE Communications Magazine* **41**, p. 92 ff, December 2003.

Mayer R.E., Models for understanding, *Review of Educational Research* **59**, p. 43 ff., 1989.

McDysan D., *QoS and Traffic Management in IP and ATM Networks*, McGraw Hill, New York, USA, 2000.

Meta Object Facility (MOF) Specification, version 1.3, OMG, March 2000.

Miller J. and Mukerji J. (editors), *MDA Guide*, version 1.0.1, OMG, June 2003.

MobiLife: please see project home page at `http://www.ist-mobilife.org`, February 2006.

Mockford K., Web services architecture, *BT Technology Journal* **22**, January 2004.

MOF 2.0 IDL Specification, OMG, July 2001.

Nichols K. and Carpenter B., *Definition of Differentiated Services Per-Domain Behaviours and Rules for Their Specification*, RFC 3086, IETF, April 2001.

OASIS: please see home page at `http://www.oasis-open.org`, February 2006.

Service Oriented Architecture Reference Model, working draft 07, OASIS, May 2005.

The NGOSS Lifecycle and Methodology, version 1.3, TMF, November 2004.

The NGOSS Technology-Neutral Architecture, version 4.1, TMF, August 2004.

The Oxford English Reference Dictionary, Oxford University Press, Oxford, England, 1995.

The Workflow Reference Model, issue 1.1, WfMC, January 1995.

Object Management Group: please see website at `http://www.omg.org`, February 2006.

OSS through Java as an implementation of NGOSS — a White Paper, TMF and OSS/J, April 2004.

Padhye J., Firoiu V., Towsley D., and Kurose J., Modeling TCP Reno performance, *IEEE/ACM Transactions in Networking* **8**, p. 133 ff., 2000.

Papazoglou M.P. and Georgapoulos D., Service-oriented computing, *Communications of the ACM* **46**, p. 25 ff., October 2003.

Parsons J., An information model based on classification theory, *Management Science* **42**, p. 1437 ff., 1996.

Personal router whitepaper, `http://wireless.ittoolbox.com/documents/academic-articles/the-personal-router-whitepaper-1395`, February 2006.

Poikselkä M., Mayer G., Khartabil H., and Niemi A., *The IMS - IP Multimedia Concepts and Services in the Mobile Domain*, John Wiley & Sons, Chichester, England, 2004.

Räisänen V., *Implementing Service Quality in IP Networks*, John Wiley & Sons, Chichester, England, 2003a.

Räisänen V., On end-to-end analysis of packet loss, *Computer Communications* **26**, p. 1693 ff., 2003b.

Räisänen V., Service quality support — an overview, *Computer Communications* **24**, p. 1539 ff., 2004.

Räisänen V., *A framework for service quality management*, submitted, 2005.

Räisänen V., Kellerer W., Hölttä P., Karasti O., and Heikkinen S., *Service management evolution*, in *Proceedings of IST summit*, Dresden, Germany, June 2005.

Ruutu J. and Kilkki K., Simple integrated media access – a comprehensive service for the future Internet, in *Proc. IFIP Performance and Communication Systems*, Lund, Sweden, May 1998.

Schneier B., *Applied Cryptography*, John Wiley & Sons, New York, USA, 1996.

Semret N., Liao R.R-F., Campbell A.T., and Lazar A.A., Pricing provisioning and peering dynamic markets for differentiated Internet services and implications for network interconnections, *IEEE Journal of Selected Areas of Communications* **18**, p. 2499 ff., 2000.

Service Framework, GB 924, version 1.9, TMF, December 2004.

Services Over IP Business Requirements, version 1.2, TMF, March 2005.

Shared Information/Data (SID) Model, version 4.5, TMF 516, TMF, November 2004.

SLA Management Handbook, version 1.5, GB 917, TMF, June 2001.

Strostroup B., *C++ Programming Language*, 3rd edition, Addison-Wesley, Reading, USA, 1997.

Tafazolli R. (editor), *Technologies for the Wireless Future*, John Wiley & Sons, Chichester, England, 2004.

TMF: please see TMF website at `http://www.tmforum.org`, February 2006.

UMA: please see UMA website at `http://www.umatechnology.org`, February 2006.

Unified Modelling Language Specification, version 1.5, OMG, March 2003.

W3C: please see home page at `http://www.w3c.org`, February 2006.

WfMC: please see coalition home page at `http://www.wfmc.org`, February 2006.

WINNER: please see project home page at `http://www.ist-winner.org`, February 2006.

Wireless Service Measurements Handbook, version 3.0, GB923, TMF, March 2004.

WS-I: please see home page at `http://www.ws-i.org`, February 2006.

WWRF: please see project home page at `http://www.wireless-world-research.org`, February 2006.

XML Metadata Interchange (XMI) Specification, version 2.0, OMG, May 2002.

Zuidweg M., Filho J.G.P., and van Sinderen M., Using P3P in web services-based context-aware application platform, in *Proc. W3C Workshop on the Long Term Future of P3P and Enterprise Privacy Languages 2003*, Kiel, Germany, 2003.

Index